Global Internet Governance

"Putting aside all the technical jargon regarding the internet, the key question that worries governments, civil society and the citizens of the world generally is the question of its governance. This volume addresses this concern head-on. Its focus on the experiences of Malaysia and Singapore – two prominent Southeast Asian countries inextricably tied up with the technology – provides fresh insights into the debates surrounding internet oversight, accountability and legitimacy. This well-researched and theoretically-informed volume by prolific and experienced media academics, Susan Leong and Terence Lee, is timely, when the world is reeling from the effects of the COVID-19 pandemic and the (in)accuracy of information provided. A time, indeed, when the governance of the internet is vital."

—Professor Zaharom Nain, *Nottingham University, Malaysia*

"This book is a useful guide for countries caught in the middle of our contemporary information technology trade war. It outlines the history of two nations – Malaysia and Singapore – that escaped the binary trap of US-China Internet Governance: US-led multistakeholderism vs China-backed multilateralism. It proposes a hybrid model that is glocal, adaptive, and concurrently neoliberal and authoritarian. Susan Leong and Terence Lee review the historical situations and practices of Malaysia and Singapore over three decades by following a third path, a hybrid internet governance model, that seems particularly well-suited for a global post-pandemic future."

—Associate Professor Weiyu Zhang, *National University of Singapore*

"Internet governance remains in crisis, without a clear roadmap for the future — so where do we turn? This brilliant and timely new book by Susan Leong and Terence Lee urges us to look beyond the fixation with US or China, or multistakeholderism or multilateralism as the default options. They propose a 'hybrid model' as the way forward, showing how this has unfolded, warts and all, in the dynamic Southeast Asian region in the cases of Singapore and Malaysia. Their rich and persuasive account underscores the importance of understanding actually-existing Internet governance as the foundation to decolonising, debugging, and reforming Internet governance for all. This book is indispensable reading for anyone concerned about the crossroads in communication and technology governance and policy today."

—Professor Gerard Goggin, *Nanyang Technological University, Singapore*

Susan Leong · Terence Lee

Global Internet Governance

Influences from Malaysia and Singapore

Susan Leong
School of Arts and Social Sciences
Monash University Malaysia
Bandar Sunway, Selangor, Malaysia

Terence Lee
College of Arts, Business, Law and
Social Sciences
Murdoch University
Murdoch, WA, Australia

ISBN 978-981-15-9923-1   ISBN 978-981-15-9924-8   (eBook)
https://doi.org/10.1007/978-981-15-9924-8

Cover illustration: © John Rawsterne/patternhead.com

This Palgrave Pivot imprint is published by the registered company Springer Nature
Singapore Pte Ltd.
The registered company address is: 152 Beach Road, #21-01/04 Gateway East, Singapore
189721, Singapore

To my beloved mother, Elsie Woo (1939–2018)
—Susan Leong

To the memory of my father, Eric Lee (1943–2019)
—Terence Lee

Acknowledgments

None of this would have been possible without the many fruitful discussions and snatched conversations along corridors with colleagues in Australia, Singapore and Malaysia. The authors wish specially to thank Ang Peng Hwa, Tania Lim, Emma Baulch, Zhang Weiyu, Michael Keane, Zaharom Nain and Helen Nesadurai for their collegiality and support.

We would also like to express our gratitude to the School of Arts and Social Science at Monash University of Malaysia for funding parts of the research work that went into this project.

Finally, to our family and friends who, as always, provide much of the most important day-to-day encouragement, words cannot fully express the thanks we have for your presence in our lives.

The later stages of this book's development took place in the shadow of the COVID-19 pandemic. Without the use of the Internet, we would have been hard pushed to collaborate as efficiently across Perth and Kuala Lumpur as we did. In this sense, then, this book is as much a product of the interoperability of the Internet as it is about global Internet governance.

ACKNOWLEDGMENTS

CONTENTS

1   Introduction—A Pair of Governance Models or More?   1
    Bibliography   9

2   The Internet in Malaysia (1990–2008): Visions
    of Technological Splendour   13
    The Multimedia Super Corridor (MSC) to Open Dissent 18
    Confidence Unfolding/Unfurling   21
    Hybridising   25
    Bibliography   28

3   The Internet in Singapore: From 'Intelligent Island'
    to 'Smart Nation'   31
    Introduction   32
    *Internet Development in Singapore: The Historical, Political
    and Ideological*   34
    The 'Intelligent' Era: 1990s to 2010s   37
    The Digital Era: iN2015 to 'Smart Nation'   42
    Digital Smarts   44
    Conclusion   47
    Bibliography   48

4   Internet Governance—The Malaysia Way   51
    If a Crime Is Committed ...   53

China's Courtship and Anti-Westernism Redux   56
Internet Sovereignty   60
Private Platforms, Public Oversight   62
Bibliography   66

5   Internet Governance: Singapore's Regulatory Influence   71
Introduction: Singapore's Governance Setting   72
The Light-Touch Principle   75
Auto-Regulation: Disciplining the Internet   81
Conclusion: The Local Is Global   85
Bibliography   88

6   Towards a Hybrid Understanding of Internet
Governance: Some Concluding Thoughts   91
Introduction   92
Glocalisation: Hybridising the Global   95
Conclusion   97
Bibliography   100

Index   103

Introduction—A Pair of Governance Models or More?

Abstract This introductory chapter frames the background to the current debate on internet governance, which consists of intersecting developments and historical conditions. It looks at the legacy of the multistakeholder model that was developed alongside the co-invention of the Internet by Anglo-American scientists, technologists and enthusiasts from the 1960s. It accounts for the unexpected mass take-up of the internet across non-English speaking and non-democratic spheres from the mid-1990s. It then brings to bear China's emphatic support for the multilateral model of internet governance since 2010. Although many of the Internet's earliest adopters envisioned it being a democratising force for democratisation, they failed to realise that the internet's spread could equally have the opposite effect of curbing liberal discourses and diminishing humanitarian values. It sets the scene to examine how two of the smaller Southeast Asian countries, Malaysia and Singapore, have governed the Internet in their respective domestic spaces over the past three decades, and what influences and lessons they hold for global internet governance.

Keywords Interoperability · Scale · Standards · Techno-nationalism · Hybrid configurations

10 years the top 10 internet companies were American. Today, 6 of the top 10 are Chinese. And this is about what kind of internet we want. Do we want an internet that's based on the values of free expression or where the government gets to decide what's allowed to be said? You know, do we want an internet where the individuals are the players or where the government itself is a major player? … America has de facto run the internet and it's based on American values because the companies were American and if that shifts to China, we will have a very different internet. . .What comes behind breaking up the US tech companies is not another US tech company right now. What clearly comes behind if you look at the numbers and the usage and the support they have, are Chinese companies. And that's a decision we have to make as a company, as a society.

    Sheryl Sandberg, COO, Facebook.[1]

On the face of it, Sandberg seems to have summed up the current state of the debate over internet governance (henceforth, IG). Do we maintain, on the one hand, that the internet should continue as it is with the individual right of freedom of expression enshrined as a guiding principle or can we countenance, on the other hand, a future where national governments determine how the internet is governed? The choice looks stark and for many the chasm between multistakeholderism and multilaterialism or what Nazli and Clark dub the distributed U.S. and the centrist China approach (Nazli and Clark 2013), makes the choice a foregone conclusion (Bradshaw and Denardis 2018; Carr 2015; Mueller 2017). Apart from those who for one reason or another are without choice, there are few adults who would opt to be more rather than less restricted in their use of a system of communication, trade and exchange than need be. The free world must rise to defend its liberalism against the creeping "neo-authoritarian" credo of Chinese communism (Jiang 2013). Indeed, as Singer and Friedman write, "many are now framing the US-Chinese relationship in cyberspace as a digital echo of that between the United States and USSR during the Cold War" (Singer and Friedman 2014). Whatever their stance what scholars of IG agree on is this: the internet is now so integrated into our daily lives (in the developing and developed world) and hence, socially, culturally, politically and economically influential that to leave the question of how the system(s) is governed

[1] A Conversation with Sheryl Sandberg, *The Future of the Internet*. Aired on Bloomberg Channel, 15 December 2019. https://www.bloomberg.com/news/videos/2019-11-27/a-conversation-with-sheryl-sandberg-video (accessed 6 January 2020).

to goodwill, ad hoc software engineering and make-do protocols is akin to misplaced trust. To decide on how the internet is governed we need, to paraphrase Berners-Lee, inventor of the World Wide Web, to "ensure that the society we build with the [Internet] is of the sort we intend" (Berners-Lee and Fischetti 1999, p. 133). To not decide is to leave the continuing function of the Internet to the "relatively flat, open, sprawling, ad hoc, and amorphous" structure that has worked so far (Bygrave and Michaelsen 2009, p. 92).

Where, you ask, is the harm in that? If the Internet had retained its modest size as an intertwined network of networks, shared between the military, engineers, scientists and academics in the Anglocentric world, rather than global network it is now, the task of IG would not have needed to stretch to much more than addressing "technical design and administrative issues" (DeNardis 2014, p. 20). The problem of how to go about IG begins, then, with an issue of scale. With the masses of data we generate, create, consume and circulate daily via the Internet, a lack of genuine oversight over how the data – personal, public and shared – is handled leaves individuals, communities, nations and a significant portion of the world vulnerable to mischief, manipulation and harm (Gillespie 2018) as demonstrated by the revelations about Cambridge Analytica (Dyer-Witheford and Matviyenko 2019, p. 1) and, those of Edward Snowden and Chelsea Manning (Van Cleave 2013). Data breaches and compromised networks are what cybersecurity fears and measures seek to eradicate. The European Union's General Data Protection Regulation (GDPR) enacted in 2018 is one initiative devised to guard the privacy of Europe's citizens.[2] At the very least, the GDPR should deter mishandling of personal data yet as the EU's own survey of small businesses found recently the gap between awareness and knowledge of the regulation, and its enforcement is wider than imagined.[3] The EU might have put a check on the compromise of customer and user data by big businesses (Kazzini 2019), still, in practice, what would be the effect of businesses separating the personal data of EU citizens from that of others? The experiment conducted and featured by the *New York Times* in 2018 (Singer and Prashant 2018) demonstrated the gulf between the amount of personal

[2] General Data Protection Regulation (GDPR) Compliance Guidelines. (2020). https://gdpr.eu/ (accessed 6 January 2020).

[3] GDPR Small Business Survey. (2019). https://gdpr.eu/wp-content/uploads/2019/05/2019-GDPR.EU-Small-Business-Survey.pdf (accessed 6 January 2020).

data on themselves individuals can obtain from digital platforms Amazon, Facebook, Google, LinkedIn, Twitter, their mobile providers and digital marketing analytics companies from the UK and the US. Microsoft has gotten round the problem by placing customers in control of their own data through a privacy dashboard (Brill 2018). The next issue that arises is how GDPR compliance might affect the interoperability of the Internet.

Interoperability is the mantra of every engineer whose designs are intended to enable the flow, be that of people, electricity and information. To compromise compatibility between the various parts and layers, in both soft and hardware terms, is to place the seamless function of the somewhat haphazardly governed Internet in jeopardy. Broadly speaking, interoperabilty rests on all parties concerned agreeing upon, setting and adhering to said standards to ensure differences are smoothed out. Transport and communication technologies were among the earliest beneficiaries of interoperability. From the first transcontinental railways (Harding 1845) and radio systems to telephone networks, interoperability hinges on faithful application of standards. Interoperability facilitates scaling. What is often obscured when standards are established, is the advantage accrued to whoever leads and establishes the standard. These days most technical standards are set by corporations based in nation-states rather than individuals so the advantage of being the originator also extends to their country of origin. This is why the U.S. government is adamant that 5G (fifth generational) technology developed by China's Huawei should not be established as the standard. Putting aside the generational leaps 5G would bring to all digital networks and users, the economic advantages of being the standard setter are enormous. As Fägersten and Rühlig (2019, p. 3) from the Swedish Institute of International Affairs observe:

> Technical standard setting might appear to be a consensual search for the technically most appropriate solution leading to absolute gains such as lower transaction costs, more efficient markets and subsequent economic growth. After all, the interoperability of products should facilitate economic growth and trade. In contemporary world affairs, however, technical standardization is more and more turning into a crucial arena for political and commercial conflict.

Whipping up concerns that wiretaps typically built into technical systems to satisfy U.S. law enforcement requirements would give Huawei (and hence, China) freedom to snoop into other countries' affairs (Johnson and Groll 2019), the U.S. government has convinced Australia, New Zealand (McDuling 2018) and Japan (Denyer 2018) to join them in banning Huawei—the world's largest telecommunications company—from supplying equipment in their countries' 5G networks. Altogether the alarm over cybersecurity and 5G standards is quite easily read as proof of the "tacit, yet possibly unintentional operation of colonial logics in certain views about internet governance" (Syed Mustafa Ali 2019, p. 112).

According to Radu, we are at the fourth stage of IG as three stages have already been and gone, beginning with "the early days of the internet, dominated by informal governance and governance through technical standards (1969-1994); followed by the period between 1995 and 2004 which saw "the boom of the commercial internet, with private actors and business practices flourishing" and then "the decade [from 2005 to 2015] of global regulatory arrangements, featuring hybrid configurations" (Radu 2019, p. 12). Certainly, a growing awareness of the influence of digital platforms has seen recent works on IG more concerned with platform governance (Gorwa 2019), and the establishment of a global digital constitution to rule the Internet (Suzor 2019) as bulwarks against the balkanisation of the Internet (Malcomson 2016). To be sure, platforms need to shoulder some of the responsibility for how their products or services are used and a global digital constitution sounds like an excellent solution to a vexing impasse. Still, if we go down that path, we also collectively gloss over the socio-cultural distinctions that make IG the complex business it is and leave most of the global South out of the equation. Why? To begin with, for most internet users outside of China, the major digital platforms are owned and operated by the developed (and Westernised) world and the universal, humanitarian values that might answer to and underpin a new digital constitution would be derived from Western mores. Not that China's digital platforms are short of success outside the nation or that parochial ethics cannot overlap with universal values but just as being the host of a dinner is different to being invited as a guest, however honoured, no IG framework can be considered genuinely global without the equal presence of all parties at the table as hosts. Historians of IG will no doubt point to the many organisations and groups such as the IETF (Internet Engineering Task Force), IAB(Internet Architecture Board), IESG (Internet Engineering Steering Group) and Internet Research Task Force (IRTF), ISOC (Internet Society) and the

IGF (Internet Governance Forum) and WSIS (World Summit on the Information Society) whose endeavours are exactly that. Other initiatives such as NWICO (New World Information and Communication Order), Net Mundial and WIF (World Internet Forum) have also emerged from the global South to participate in this important conversation. However, as the decline of WSIS, Net Mundial and NWICO show, worldwide consensus is a herculean task (Hubbard, and Bygrave 2009). Covering the establishment and demise of the various initiatives is beyond the limited scope of this volume. We argue, instead, for a pause before we hurtle into the fourth stage of IG to gather the significant value to be found in an examination of the hybrid configurations enacted and experienced in Malaysia and Singapore.

Bygrave and Bing identify five different ideal types or models of IG beginning with "(*a*) the model of spontaneous ordering; (*b*) the model of transnational and international governance institutions; (*c*) the model of code; (*d*) the model of national regulation; and (*e*) the model of market-based ordering (Bygrave and Bing 2009, p. 55). We posit that combinations of these ideal types have successfully been used in Malaysia and Singapore since the introduction of the Internet to these two countries and, that an exploration of what has transpired, and why, will yield knowledge that could enlighten what Radu calls the fourth stage of IG. With populations of 32 and 5 million, respectively, neither Malaysia or Singapore are especially large or prominent on the world stage although the latter's economic prowess has been the object of some admiration. Whereas Malaysia's experimental approach has produced some unexpected fruit (Leong 2014), Singapore's technocratic approach has been much maligned even in the early days of the Internet (Barlow 1996) and continues to be the target of much criticism (Lee 2010). What is of interest here is how, as part of Southeast Asia, both nations remain the theatre where struggles between the super powers manifest as daily decisions, actions and consequences by and for both states and citizens. Not unusually for that period in history, Malaysia and Singapore obtained independence from the British between the 1950s and 1960s, and went on to establish themselves as sovereign nation-states. Forced or freed, depending on one's perspective, to manage on their own, both the Malaysian and Singaporean governments have employed techno-nationalism within their social imaginaries to inspire harmony, diligence and prosperity. Such are the conditions that greeted the Internet in the mid-1990s when it first arrived in both countries. Such are also the conditions that give rise to hybrid formations developed rapidly, sometimes expediently. Experimental and technocratic, few would be able to point to two more vastly divergent approaches toward the early Internet within such a small circle and similar national foundations. More recently, these differences appear to have been erased. Notably, when China's march toward cyber-supremacy has been in question. For example, neither Malaysia or Singapore have banned Huawei from the development of 5G networks in their countries. Malaysia, in particular, has been seen to endorse Huawei by Prime Minister Mahathir's visit to their factory in Beijing within the first 12 months of taking office again (Jin and Soo 2019). Though their beginnings are similar enough, Singapore having left the Federation of Malaya just two years after it was formed in 1965, both nations have fared rather differently so that earnings per capita in Singapore today are among the highest in the world at USD 94,679,[4] while earnings per capita in Malaysia are at USD 30,650.[5] There is no direct correlation between how nations approach IG and their national fortunes per se as myriad factors go into how well countries do and by what measures. However, considering the pivotal role technologies play in the task of nation-building and the newly industrialised economies (NIEs) of Southeast Asia (Leong 2014), there can be no illusion that some impact ensues from what countries imagined and expected of the Internet and treated it introduction into the region in the mid-1990s. It might seem odd to think of Singapore at the same time as we speak of the Global South. The figures cited in previous pages argue against the city-state's inclusion in any such formation but when Singapore first achieved nationhood it was far from certain that the country would survive. So it is appropriate to trace the course of hybrid IG formations via Singapore as well as Malaysia.

It is important to understand the historical difference between internet governance and internet regulation. The former is the broader structure of conventions, both technical and legal, upon which the latter operates. This difference is why content can be freely hosted and available online yet

---

[4] The World Bank. (2019). *World Development Indicators.* https://databank.worldbank.org/reports.aspx?source=2&country=SGP (accessed 4 January 2020).

[5] The World Bank. (2019). *World Development Indicators.* https://databank.worldbank.org/reports.aspx?source=2&country=MYS (accessed 4 January 2020).

blocked for users in a specific country. While the governance and the regulation of the internet are necessarily intertwined, governance has global implications whereas regulation can sometimes be a concern only for the governments of individual nation-states. In practice today, the distinctions between the governance of the Internet and the regulations surrounding the Internet are seldom as clear cut. In this volume, as a reflection of the multitude of changes that the concept of IG has undergone, we define IG as the underlying combination of technological conventions, regulations and systems of management that enable the working of the Internet. In doing so, our intent is to undo some of the masking of an "extant, incursive, hegemonically 'Northern' (that is, 'Western', West-centric etc.) system of internet governance" (Syed Mustafa Ali 2019, p. 110) that passes for global internet governance today.

Hindsight allows us to see how, depending on the perspective taken, Malaysia's responses could be viewed as overestimating (economic) or underestimating (socio-political) the implications of introducing the Internet to society. Whereas Singapore's responses might be seen as just right (economically) and over-engineered (socio-political). It is worth pointing out that until May 2018, Malaysia, had been governed by the same coalition of political parties for over six decades. The flattened hierarchy of internet-enabled messaging apps is said to have tilted the scales in favour of the then opposition and vital to the election of a new government (Leong 2019). Yet, Singapore with its intelligent island policies, world class infrastructure and affluent, highly-literate citizenry continues to be governed by the same party that has held power since independence. The hybrid IG configurations each country assembled, adapted and continue to tweak speak firstly, to the political economic situations they faced when then new and relatively unknown suite of technologies arrived on their shores; and secondly, to their socio-cultural interpretations of what was variously hyped as the "Information Superhighway",[6] "the network of networks" and cyberspace.[7]

---

[6] Al Gore Speech, Information Superhighways, International Telecommunications Union, 21 March 1994: AMDOCS: Documents For The Study Of American History. (1994) http://vlib.iue.it/history/internet/algorespeech.html (accessed 5 January 2020).

[7] WIRED. (2009). *March 17, 1948: William Gibson, Father of Cyberspace.* https://www.wired.com/2009/03/march-17-1948-william-gibson-father-of-cyberspace-2/ (accessed 5 January 2020).

Hence, in the five chapters that follow we explore the paths leading up to the current status quo for both Malaysia and Singapore, with two chapters devoted to each country. The chapters are arranged in a chronological fashion to facilitate the contrast and comparison between each country's chosen paths and their unfolding. Broadly speaking we take a socio-historical approach in this volume, reading the hybrid formations of IG as socio-technical outcomes. Due to constraints of space, the periods we cover here range from the 1990s to the present day. In the fifth chapter, we discuss the lessons that can be drawn from the emergence of the hybrid IG formations of Malaysia and Singapore and what influence their experience could have on the enduring debate over how the Internet should/can be governed. The hope is to dislodge the tacit assumption, intentional or not, that "the West is the location of the Internet and the Rest becomes the side to be rescued by the Internet" (Shah 2017, p. 50).

BIBLIOGRAPHY

Barlow, J. P. (1996). *A Declaration of the Independence of Cyberspace*. https://www.eff.org/cyberspace-independence. Accessed 4 January 2020.

Berners-Lee, T., & Fischetti, M. (1999). *Weaving the Web*. London: Orion Business.

Bradshaw, S., & Denardis, L. (2018). The Politicization of the Internet's Domain Name System: Implications for Internet Security, Universality, and Freedom. *New Media & Society, 20*(1), 332–350.

Brill, J. (2018). *Microsoft's Commitment to GDPR, Privacy and Putting Customers in Control of Their Own Data*. https://blogs.microsoft.com/on-the-issues/2018/05/21/microsofts-commitment-to-gdpr-privacy-and-putting-customers-in-control-of-their-own-data/. Accessed 6 January 2020.

Bygrave, L. A., & Michaelsen, T. (2009). Governors Of Internet. In L. A. Bygrave & J. Bing (Eds.), *Internet Governance* (pp. 92–105). Oxford: Oxford University Press.

Bygrave, L. A., & Bing, J. (2009). *Internet Governance*. Oxford: Oxford University Press.

Carr, M. (2015). Power Plays in Global Internet Governance. *Millennium—Journal of International Studies, 43*(2), 640–659.

DeNardis, L. (2014). *The Global War for IG*. New Haven: Yale University Press.

Denyer, S. (2018). *Japan Effectively Bans China'S Huawei And ZTE From Government Contracts, Joining U.S.* https://www.washingtonpost.com/world/asia_pacific/japan-effectively-bans-chinas-huawei-zte-from-government-contracts-joining-us/2018/12/10/748fe98a-fc69-11e8-ba87-8c7facdf6739_story.html. Accessed 5 January 2020.

Dyer-Witheford, N., & Matviyenko, S. (2019). *Cyberwar and Revolution: Digital Subterfuge in Global Capitalism.* Minneapolis and London: University of Minnesota Press.

Fägersten, B., & Rühlig, T. (2019). *China's Standard Power and Its Geopolitical Implications for Europe.* https://www.ui.se/globalassets/ui.se-eng/pub lications/ui-publications/2019/ui-brief-no.-2-2019.pdf. Accessed 4 January 2020.

Gillespie, T. (2018). *Custodians of the Internet: Platforms, Content Moderation, and the Hidden Decisions That Shape Social Media.* London: Yale University Press.

Gorwa, R. (2019). What Is Platform Governance? *Information, Communication & Society, 22*(6), 854–871.

Harding, W. (1845). *Railways the Gauge Question: Evils of a Diversity of Gauge, and a Remedy* (2nd ed.). London: J. Weale.

Hubbard, A., & Bygrave, L. A. (2009). Internet Governance Goes Global. In *Internet Governance : Infrastructure and Institutions* (pp. 213–235). Oxford: Oxford University Press.

Jiang, M. (2013). A New Internet World, A Neo-Authoritarian Model Of Internet Governance. *China Policy Institute Blog.* http://blogs.nottingham. ac.uk/chinapolicyinstitute/2013/11/18/a-new-internet-world-a-neo-author itarian-model-of-internet-governance/. Accessed 6 January 2020.

Jin, M., & Soo, Z. (2019). Mahathir's Visit To Huawei in Beijing Seen As 'Endorsement'. https://www.scmp.com/tech/enterprises/article/3007585/ malaysian-pm-mahathir-mohamads-visit-huawei-beijing-seen-sign. Accessed 5 January 2020.

Johnson, K., & Groll, E. (2019). *The Improbable Rise of Huawei.* https://for eignpolicy.com/2019/04/03/the-improbable-rise-of-huawei-5g-global-net work-china/. Accessed 5 January 2020.

Kazzini, K. (2019). *Europe's Huge Privacy Fines Against Marriott and British Airways Are a Warning for Google and Facebook.* https://www.cnbc.com/ 2019/07/10/gdpr-fines-vs-marriott-british-air-are-a-warning-for-google-fac ebook.html. Accessed 6 January 2020.

Lee, T. (2010). *The Media, Cultural Control and Government in Singapore.* Abingdon: Routledge.

Leong, P. P. Y. (2019). *Malaysian Politics in the New Media Age Implications on the Political Communication Process.* Singapore: Springer.

Leong, S. (2014). *New Media and the Nation in Malaysia: Malaysianet.* London: Routledge.

Malcomson, S. L. (2016). *Splinternet: How Geopolitics and Commerce Are Fragmenting the World Wide Web.* London: OR Books.

McDuling, J. (2018). *New Zealand Joins Australia in Banning Huawei*. https://www.smh.com.au/business/companies/new-zealand-joins-australia-in-banning-huawei-20181128-p50iz5.html. Accessed 5 January 2020.

Mueller, M. (2017). *Will the Internet Fragment?: Sovereignty, Globalization and Cyberspace*. Oxford: Polity Press.

Nazli, C., & Clark, D. D. (2013). Who Controls Cyberspace? *Bulletin of the Atomic Scientists, 69*(5), 26–27.

Radu, R. (2019). *Negotiating IG* (1st ed.). Oxford Scholarship Online.

Shah, N. (2017). The State of the Internets: Notes for a New Historiography of Technosociality. In G. Goggin & M. McLelland (Eds.), *The Routledge Companion of Global Internet Histories*. London: Routledge.

Singer, N., & Prashant, S. R. (2018, May 20). *U.K. vs. U.S.: How Much of Your Personal Data Can You Get?*. https://www.nytimes.com/interactive/2018/05/20/technology/what-data-companies-have-on-you.html. Accessed 6 January 2020.

Singer, P. W., & Friedman, Allan. (2014). *Cybersecurity and Cyberwar: What Everyone Needs to Know*. London: Oxford University Press.

Suzor, N. P. (2019). *Lawless: The Secret Rules That Govern Our Digital Lives*. Cambridge: Cambridge University Press. https://doi.org/10.1017/9781108666428.

Syed Mustafa Ali. (2019). Prolegomenon to the Decolonization of Internet Governance. In Daniel Oppermann (Ed.), *Internet Governance in the Global South: History, Theory and Contemporary Debates*. Sao Paulo: International Relations Research Center, University of Sao Paulo.

Van Cleave, M. (2013). *Myth, Paradox & the Obligations of Leadership: Edward Snowden, Bradley Manning and the Next Leak*. Center for Security Policy.

CHAPTER 2

The Internet in Malaysia (1990–2008): Visions of Technological Splendour

Abstract This chapter addresses the crucial period between Mahathir's articulation of Vision 2020 in 1996, the Multimedia Super Corridor (MSC) opened in 1998 and, the loss in 2008 of its two-third majority in parliament by the long-standing government for the first time in history. Two facts amplified the extraordinary powers the state had to suppress dissent or criticism i.e. the lack of transparency surrounding media ownership and an unchanged government in power from 1957 to 2018. Malaysian authorities' iron grip was relaxed in 1998 with the establishment of the MSC as a technology zone. The government's promise not to censor the Internet proved to be the first in a long line of events that saw many Malaysians become active, critical and informed participants of digital media. The process whereby digital media in Malaysia went from an alternative to broadcast media and upstart challenger to become the de rigueur source of news, information and data for many is also the story of how Malaysia's hybrid model of Internet governance came to be.

Keywords MSC · Vision 2020 · Technology corridor · Bill of Guarantees

© The Author(s), under exclusive license to
Springer Nature Singapore Pte Ltd. 2021
S. Leong and T. Lee, *Global Internet Governance*,
https://doi.org/10.1007/978-981-15-9924-8_2

13

In the 1980s through to 2003 and again, since May 2018, Mohamad Mahathir was the Prime Minister (hereafter, PM) of Malaysia. Over his first 23-year term Prime Minister Mahathir has had a strong influence on the direction the nation took, both economically and culturally. Consequently, how he imagined and understood the Internet was also pivotal to its place in the task of nation-building. For reasons of space constraints, this volume's coverage begins with the 1990s. Nonetheless, a brief foray into the decades immediately preceding the 1990s is unavoidable if we are to understand why and how the IG developed the way it did in Malaysia. In 1982, Mahathir announced the 'Look East' policy (Furuoka 2007) and explained it as "emulati[on] of the rapidly developing countries of the East in the effort to develop Malaysia" and, singled out the qualities of "diligence and discipline in work, loyalty to the nation and to the enterprise of business where the worker is employed, priority of group over individual interest" (Jomo 1988, p. 10) for attention. The idea was to have a transfer of values together with that of technologies and knowledge from Japan (Furuoka, p. 509). In addition to striking a nationalist tone, Mahathir also professed a streak of anti-Westernism (Furuoka, pp. 507–508). In fact, he sought alternative development strategies (to the established British model) and role models for Malaysia from the East. While much inbound investment was, indeed, introduced into the country with the help of the 'Look East' policy, this was due in no small part, as Anazawa suggests, to the appreciation of the yen against the greenback (Anazawa, cited in Furuoka, p. 512). Vast amounts were also invested by the Mahathir government sending young Malaysians to Japan for further education. Notable success stories of the Look East Policy included the launch of the Malaysian national automobile maker, Proton, with the aid of Japan's HICOM. However, but by the end of the 1980s it was clear that the hoped-for transfer of values had failed to occur. There were many reasons but as Furuoka pointed out (2007, p. 509): "the 'Look East' policy did not take into account fundamental differences in Japan's and Malaysia's social and cultural backgrounds". While the intended objective of technology transfer succeeded to some extent, the hoped for transfer of values failed to eventuate. As will become apparent in the latter part of this chapter, the lesson surrounding the transfer of values was not lost and may, indeed, explain the approach taken when Mahathir embarked upon the next nation-building mega project. Of interest here is how, Mahathir sought even then to combine the best of the East with the best of the West to create hybrid models (Mahathir, cited in Saravanamuttu 1988, p. 7):

> Look East means we should resort to other sources than just the West and this doesn't mean that we are going to give up the West entirely. What is good in the West, we will still follow, but here is a source of ethical values, systems and everything else which are useful to us. Why shouldn't we make a deliberate effort to acquire this from the East?

Alongside the above-mentioned economic initiatives are three national policy frameworks stretching across from 1971 to 2010 that sought to redress the "persistent disparities in inter- and intra-ethnic distribution as well as differences between rural and urban incomes and between less developed and more developed regions" (PMO 2006, p. 3). These policies are: the New Economic Policy (NEP, 1971–1990), the National Development Policy (NDP, 1991–2000) and the National Vision Policy (2001–2010). While the national mission has ostensibly been couched to include broader, national ethnic and urban/rural divides (Abdulai 2004, pp. 6–7), it is primarily social inequity between the *bumiputeras* (literal translation: sons of the soil; mostly Malays but including the indigenous peoples of east and West Malaysia) and the Chinese minority. Yardsticks of measure such as "mean income ratio of Bumiputera to Chinese/Indian" featured in official documents such as *The National Mission* (PMO 2006, pp. 10–11), belie the broader focus. Officially Malaysia, like Singapore, has four races known by the acronym, CMIO, representing the Chinese, Malay, Indian and Others. The last category being a catch-all for all those who could/would not fit neatly into one of the pre-determined groups. The majority of the population is of Malay race while the Chinese, Indians and Others are minorities in order of size. The indigenous people of Peninsula Malaysia and to a great extent, Borneo Malaysia have gradually been absorbed into the *bumiputera* category. These categories were themselves imposed upon the population during the colonial era but perpetuated by the new nation-state. There are historical inter-ethnic frictions for such a specific focus that, among others, have their roots in the British colonisation and devolution that led to independence for Malaysia in 1957. Very briefly, Malaysia was established as a constitutional monarchy with 13 states, 3 federal territories and ruled by the *Barisan Nasional* (National Front) up until 2018 with the help of nine traditional rulers, sultans. A dozen years after independence, an election in 1969 that saw socio-political tensions morph into ethnic conflict, resulted in a racial riot on 13 May. The vitriol exchanged during the event and the violence perpetrated on fellow Malaysians remain a scar on the nation's imaginary and continues to disproportionately colour interactions among Malaysians (Leong 2016b). By the time Doctor Mohamad Mahathir became the fourth PM of Malaysia in 1981, the affirmative action policy embedded in the NEP, NDP and NVP over 4 decades had hardened into a perpetual expectation from the *bumiputera* and began to have greater ramifications for politics, the economy and society. The relevant point here is: however differently presented, all three comprised affirmative action policies sought to provide assistance to disadvantaged *bumiputeras* including the mandatory 30% *bumiputera* ownership of all businesses in Malaysia, business hiring practices, higher education quotas and the medium of instruction of the public education system. This was the context within which the regulations and policies enacted to enable the MSC emerged in the mid-1990s.

In the meantime, when the Look East policy failed to deliver a more enduring transfer of values at the end of the 1980s, the government in Malaysia needed a new narrative and a new set of promises to convince the citizenry that 'the good life' was still attainable. In 1991, PM Mahathir delivered a speech titled, *Vision 2020: The Way Forward* (Mahathir 1991) in Kuala Lumpur (hereafter, KL), the capital of Malaysia in which he outlined a new direction for the nation. Among other things he envisioned a future where all races in the nation would be able to thrive convivially and work together to attain developed nation status by the then distant year of 2020. At the time of writing, Malaysia is deemed a middle-income developing nation by the World Bank, whereas Singapore has been a high-income nation since 1987.[1] Even then Mahathir was adamant that (1991):

> We should be a developed nation in our mould ... Malaysia should not be developed only in the economic sense. It must be a nation that is fully developed along all the dimensions: economically, politically, socially, spiritually, psychologically and culturally. We must be fully developed in terms of nation unity and social cohesion, in terms of our economy, in terms of social justice, political stability, system of government, quality of life,

[1] The World Bank (2019).

social and spiritual values, national pride and confidence. (Prime Minister Mahathir 1991)

It was this determination to forge its own path that had led Mahathir and, hence, Malaysia first to look east and then, when the objectives of that policy did not come into fruition, to seek an alternative direction. At the point in time when Mahathir laid out Vision 2020, the cyberlibertarian ethos of the early Internet was rife in all walks of life in Western societies. In the United States the economic policies of President Ronald Reagan (Dreier 2011) during the 1980s had provoked the rise of a tech-infused counter-culture that produced individuals such as Steve Jobs and Steve Wozniak. Together, the two Steves started Apple and sold 200 hand-built units of Apple I between 1976 to 1997 (Rawlinson 2017). Their user-friendly machines saw time-sharing for access to a mainframe become a thing of the past. Other individuals, like Stewart Brand, publisher of the *Whole Earth Catalog*, proclaimed in 1984 that "information wants to be free" and the rest of the world promptly forgot that in the same breath he also added: "it also wants to be expensive" (Brand 1998). Counter culture *Wired* magazine began its first run in 1993. Among those featured on its glossy, full-coloured covers was Jeff Bezos, CEO of Amazon, who dreamt in 1994 of the untold and amazing advances (and fortunes) to be made by taking their business idea online (Amazon Startup Story 2020). Powered by collective enthusiasm the Dot Com Bubble was inflating at exponential rates. Still, John Perry Barlow, former lyricist of the band, Grateful Dead, penned the defiant "A Declaration of the Independence of Cyberspace" in 1996, setting up the battle as one between governments and their people, corporations and consumers (Barlow 1996).

From where we stand today, the exuberance that greeted the early Internet is difficult to comprehend especially if we can recall how clunky and slow to upload browsers or how web sites were generally, built by coding, not very eye-catching or sophisticated but cumbersome. The work of managing the Domain Name System was initially performed by one volunteer, Jon Postel, later through IANA (Internet Assigned Numbers Authority) in 1988 until ICANN (the Internet Corporation for Assigned Names and Numbers) took over the task in 1998 (ICANN History Project). Yet, even in Asia, the early Internet bred a kind of unbridled optimism among early adopters that was both heady and infectious. Enthusiasts and hobbyists such as, Indonesian engineer and academic, Onno W. Purbo, had repurposed the Chinese wok's parabolic curve (*Wajanbolic*) to bring low cost Internet connection to parts of the archipelago. This author's own acquaintances at the Singapore Polytechnic had been lugging around a telephone receiver joined to a wood block for months in the early 1980s. Although an earlier computer network, RangKom, had been put together in 1987 for academic exchange, it was not until 1990 when JARING (Joint Advanced Research Integrated Networking) was built that use of the internet slowly diffused into the wider Malaysian community (Shariffadeen 1994, pp. 7–8). Even if Malaysia was to become a developed nation in its own mould, the die was cast by the time the Internet was accessible to the general public in the mid-1990s.

THE MULTIMEDIA SUPER CORRIDOR (MSC) TO OPEN DISSENT

The MSC was launched in 1996 at a gala business event where the PM announced the planned direction of the venture. The physical dimensions of the MSC are well known, a specially designated 15 by 50 kilometre greenfield site south of Kuala Lumpur. In broad brushstrokes charged with what, in the context of this volume, can only be described as cyberlibertarian utopianism, Mahathir painted the portrait of a futuristic "intelligent garden city" as the new administrative capital that he dubbed Putrajaya, alongside IT City and the then yet to be completed Kuala Lumpur International Airport acting as the gateway to the MSC. The concept of the MSC was encapsulated by Mahathir as "a multicultural 'web' of mutually dependent international and Malaysian companies collaborating to deliver new products and services to customers across an economically vibrant Asia and the world" (Mahathir 1996). The ultimate objective of this mega-sized project being to create a "new codes of ethics in a shrunken world when everyone is a neighbour to everyone else, where we have to live with each other without unnecessary tension and conflicts" (Mahathir 1996). "I see", Mahathir declared, "a global community living at the leading edge of the Information Society" (Mahahtir 1996). Investors were promised "rich returns" and much pains were taken to persuade them of the MSC's viability, citing the politically stable government, lack of "volatile local politics" and, "the political will and the power to rapidly change any existing laws or policies that impede the ability of companies to capitalise on the benefits afforded by the Information Age" (Mahahtir 1996). For locals, the MSC was couched as a "test drive" and/or an experiment which would allow [only] "MSC companies the unrestricted import of knowledge workers for the next ten years. In addition, there will be no employment restrictions on MSC companies and there will be no restrictions on foreign ownership". In other words, the MSC would be a corridor of exception where the main aspects of the affirmative action policy would be suspended. The assumption implicit within this limited concession was that by treating the MSC as a technology corridor and isolated zone, any fallout from these experiments would be contained (Leong 2014). Of course, no concessions to the affirmative action policies were countenanced outside of the MSC zone. Exhaustive research, the PM assured them, had already been conducted to identify key factors for success. To "close the gap immediately", it was imperative that companies within the 750 square kilometre corridor had (Mahathir 1996):

> access to sufficiently skilled human resources and flexibility in hiring; access to world-class telecoms and information infrastructure; liberalised financial environment with *no local content/ownership/partnership requirements*; quality of life as good as home countries with every convenience and ease of doing business.

Again, in 1995 Mahathir explained that whereas the Malaysian government had "in the past... facilitated the development and application of IT through the bottom-up approach...Lately, we have decided that the bottom-up approach has to be completed by the top-down policy planning and management scheme as well" if Malaysia was to succeed with its objectives for the MSC (Mahathir 2002d, p. 235). Two years later in a speech delivered to a domestic audience in 1997 titled, Inventing Our Common Future, Mahathir spoke of the MSC as the "tentative first bridge" to the digital future. Further insight into how Mahathir and, for all intents and purposes, Malaysia, viewed and valued governance can be found in his description of it "as the 'umbrella concept embracing and defining this process of reinvention and therefore is the most important" (Mahathir 2002b). At the same time, as Mahathir himself acknowledged, Malaysia's planned transition into this digital future owed something to the influence of MIMOS (Malaysian Institute of Microelectronic Systems) and its then CEO, Tengku Mohd Azzman Shariffadeen (Mahathir 2002c). It is, therefore, most interesting to note that in 1994, Shariffadeen wrote of the state assuming the function of a "creative" regulator of technical change, despite the trend towards deregulation" with the caveat that "an overcontrolled environment will also have an adverse effect, restraining innovation by individuals and organisations" and advocated for a "proactive yet balance [sic] approach... to successful exploitation of IT". These, then, are the perspectives that formed the foundation from which Malaysia embarked on the formulation of many new regulations or Cyberlaws to facilitate the MSC's function. They include: "the Digital Signature Act, the Computer Crimes Act and the Telemedicine Act, all of 1997, the Communications and Multimedia Act of 1998, and various amendments to the Copyright Act" (Antons 2006). The Malaysian Communications and Multimedia Act 1998 (MCMA) remains the overarching act for the cyberlaws. In addition, eight flagship applications—paperless and electronic government, national multipurpose smart cards, borderless marketing, manufacturing web, research and development clusters, smart schools, multimedia financial haven and tele-medicine—were identified when the MSC mega project was launched ('People-Oriented' IT Agenda Required 1996). The idea of a Cyber Court of Justice was also mooted, one where Malaysia hoped to "develop and influence the new culture of global information" so as to balance "the conflict between control and licence and offsetting the adverse social effects of a massive unremitting onslaught of instant, uncensored information" (2002a, p. 140). In 1996 the government also devised the MSC Malaysia Bill of Guarantees (BoGs) to ensure "the best tailored incentives and financial/venture capital environment for investors" and "provide MSC with the best operating environment in Asia" ('Bill to ensure best deal for investors'). While there are many worthy and important tax, employment and infrastructural promises contained in the BoGs, it was and is on the seventh that much of what later transpired hinged i.e. "to ensure no censorship of the internet" ('MSC Malaysia Bill of Guarantees').

It is important to note that with the MSC, Mahathir's anti-Westernism went into partial abeyance even though he sought still to create a hybrid economic technological model rather than a clone of America's Silicon Valley. In fact, in almost all of the speeches delivered to foreign investors in various cities across the world, Mahathir repeatedly emphasised that "the MSC is not just a physical location...and it is not a far eastern imitation of the Silicon Valley ... The MSC is envisioned to be a high tech testbed that will allow companies to explore multimedia technologies without any limitations" (Mahathir 2002a, p. 107). There was, I think, no lack of foresight in how the internet would change Malaysian society. For example, Mahathir cautioned very early on that, "it is important that we use ICT wisely, according to our values and culture... the only way to ensure both access and appropriateness is to take charge of technology development ourselves" (Mahathir 2002c, p. 91). However, as events attest, the utopianism and miscalculation obscured much of the potential scale of change and impact when Internet users are gripped by intense dissent such as that wrought by *Reformasi*. Particularly when juxtaposed against the control exerted over broadcast media in the same national space/imaginary. It is for this reason that I have elsewhere, referred to the implications of the Internet' in Malaysia as incongruous and dissonant (Leong 2014).

Confidence Unfolding/Unfurling

Through a confluence of events, timing and the imaginaries of the Internet invoked, the absence of censorship on the Internet became the start of somewhat unexpected socio-political outcomes for the country. The containment of the BoGs' slate of incentives, right and privileges to those businesses and users physically within the MSC was, with the benefit of hindsight, always going to be a tough proposition. Although Singapore had managed to erect a *cordon sanitaire* around the island-state's Internet space, Malaysia was altogether bigger geographically and more heterogeneous socio-politically with pockets of the country lagging behind in their development. Consequently, the promise of non-censorship created a loophole that was quickly exploited by locals. Why such enthusiasm? Over the course of history from 1511 to 1957, Malaysia had been colonised by many European countries but it was the British, the last of the colonisers who left the country the legacy of a media policy and regulations framework, heavily inflected by the desire to quell the communist threat and insurgents. Broadcast media outlets, for example, were required to renew licenses annually at the discretion of the Minister of Home Affairs and sedition was punishable by detention without trial (Leong 2016a). The bulk of the mainstream outlets were either owned and managed by the elite heavily invested in the continued rule of the *Barisan Nasional* (Hopkins 2014). Understood and imagined as an industrial technology, albeit one digitised and capable of "magic" (Mahathir 1996), the potential of the Internet as a medium of communication was underestimated. With the government's promise not to censor the Internet in Malaysia ringing in their ears and the presumably lower costs of setting up digitally, online media outlets such as *Malaysiakini*, *The Malaysian Insider*, *Merdeka Review*, *The Edge* and *The Nut Graph* were established and for some time at least between 1998 to 2007 thrived. Citizens also became progressively bolder in the content they produced and shared on self-built websites. Alternate media became an idea with substance to it in Malaysia. Alternate is, of course, used here in reference to the anaemic mainstream and broadcasting media, ever prepared to toe the party line up until 2018 when the coalition lost the election. Dissent from many quarters and on many issues have always been a part of Malaysian civil society but with the help of the Internet many more avenues to disseminate alternate views became popular to ordinary Malaysians. The search for other information sources was helped along when smartphones became part and parcel of everyday life from 2007 onwards and, infrastructural/technological advancements introduced first 3G, then 4G, Wifi, broadband, mobile telephony and mobile Internet to the country.

The political events that shook the nation from the 1990s to 2008 also gave impetus to further increases in internet use. The Deputy Prime Minister of Malaysia in the late 1990s was Anwar Ibrahim (hereafter, Anwar). With a strong scholarly Islamic background, he was popular among the many Malays and cut a charismatic and internationally well-known figure. His shock dismissal from his post in 1 September 1998 (Khoo 2003, p. 72), arrest for corruption and rough treatment during detention was widely perceived as unjust (p. 100). It also created a wedge within the Malay community that saw public outrage become a major catalyst of public dissent. Both supporters and detractors aired their views freely online (Leong 2019, p. 7) on sites like *Harakah* and *Reformasi Diary* alongside Malay language magazines such as *Harakah*, *Detik*, *Tamadun*, *Wasilah* and *Eksklusif* (Funston 2000; Khoo 2003). The resultant *Reformasi* movement was a major turning point for all Malaysians but especially pivotal period for internet users in Malaysia as it pushed BoG 7 to the limit. Khoo's description of the unfolding dilemma posed by *Reformasi* for the Mahathir administration is worth repeating here:

> Overnight, Reformasi brought an unintended fulfilment of the regime's slogan, *Cintai IT!* All sorts of Reformasi-minded individuals 'loved IT', and especially the Internet, for enabling them to post information, access materials and connect with other people in ways that were free of state censorship, even if the users were not quite liberated from fears of

state retribution. Here, the regime's promotion of the Multimedia Super Corridor (MSC) turned out to be a boon for its critics. The information superhighway was difficult to police as the regime could not tamper with or shut down websites without violating the freedom of expression promised in the MSC's 10-point 'Bill of Guarantees that Mahathir had offered international investors.

Hobbled by the undertaking of BoG 7, the government was unable to do very much. Historically, media governance in Malaysia has based primarily on the broadcast model, i.e. one-to-many transmission. So broadcasters, publishers and media distributors of all kinds were subordinated to state agendas because their licenses to operate whether by securing spectrum or license to circulate information were issued by national governments under the Printing Presses and Publication Act 1984. In short, all purveyors of information were subject to having their operations shut down if licenses were not renewed and, up until 2012 the PPA in Malaysia stipulated that all media licenses be renewed annually at the discretion of the Home Minister without judicial review (Leong 2016a, pp. 3–4). Since online news sites are not officially categorised as newspapers and, therefore, not subject to the strictures of the PPPA. This made it relatively difficult to police alternate and online sources of news and information. Without the instruments of media governance afforded by the PPA, successive governments led by Abdullah Badawi (2003–2009), Najib Razak (2008–2018) that followed after Mahathir first retired in 2003, resorted to other parts of Malaysia's penal code and at various times, used the Internal Security Act (Leong 2019, p. 8), the Sedition Act (Leong 2019, p. 123) and Section 233(1)(a) of the Communications and Multimedia Act 1998 to silence detractors. The premises of independent online news site, *Malaysiakini*, for example, was raided and/or investigated by the police for sedition in 2003 and again, in 2015 for slander ('Police raid offices of Malaysiakini.com News Website' 2016).

In late 2003 PM Mahathir retired leaving then Deputy Prime Minister Abdullah Badawi (hereafter, Badawi) to step into the role of Prime Minister in 2004. PM Badawi had a softer and more relaxed approach towards matter but, even so, did introduce amendments to The Election Act and Election Offences Act to include a new code of ethics for political campaigns (Leong 2019, p. 88). Mobile phones, short messages (SMS), websites and emails were used by both the incumbent ruling coalition (BN) and the opposition. The landslide victory in Malaysia's 11 General

Election secured by BN appeared to endorse the Badawi administration. According to Pauline Leong, for a few years after there was a "freer democratic space, especially for media" (Leong 2019, p. 90). Notably, this was also the period, 2004–2008, when the Badawi administration introduced the idea of *Islam Hadari* (Civilisational Islam) as a form of nation-building. A moderate Islam "aimed at reminding Muslims that Islam did not prevent them from engaging with Modernity, pluralism, diversity, and the challenges of the modern age" (Noor 2013, p. 93), the idea of *Islam Hadari* in Malaysia quickly became mired in politicisation and ultimately, failed to gain traction.

Meanwhile, the general public, now better apprised of information on Malaysian matters, political, social and economic with broader sources via the Internet were even more emboldened and equipped to contest official narratives. This took the form of incidents that saw sections of the populace rise up in defiance, helped along by the use of the Internet and instant messaging to engage and mobilise others. The Hindu Rights Action Force rally, for example, mobilised many aggrieved Malaysian Indians in 2007, to hold a major protest in protest at the poor treatment meted out to their community (Leong 2009). On 25 November 2007, thousands of ethnic Indians in Malaysia marched to the British High Commission to hand in a petition for their historical dispossession in Malaysia to be addressed by the ex-colonisers. The rally was preceded by a series of videos, photographs and polemics blog entries circulated online, of Hindu temple demolitions by authories and unilateral conversions of children by one parent to Islam (Leong 2009). The Hindraf Rally was preceded by another held on 10 November in the same year. The first Bersih (clean) rally was organised by a coalition of non-government organisations to demand for clean and fair elections. With prominent lawyer-advocate and President of the Malaysian Bar Council, Ambiga Sreenevasan, at its helm together with a host of other human rights activists and bodies, Bersih had crafted a secular approach that shifted the focus away from race politics to the disenfranchisement of Malaysians. Its impact was unexpectedly large especially in a country where citizens were habitually silenced in the name of national unity. In a manner typical of semi-authoritarian nations and somewhat incredulously, the government brought out the heavy equipment, firing tear gas and aiming water cannons at its own citizens. In any event, it is possible to see how the government's response might well have spurred anti-government sentiments, hardened the resolve of the protestors and provoked even those new to dissent to express their objections.

In the 12th General Election held in 2008, the BN coalition lost its two-thirds majority in parliament. Held since independence by the BN coalition and regarded as conferment of the mandate rule as it saw fit, the loss of the majority was a blow to the government. Consequently, PM Badawi bowed to pressure from his own party (UMNO) and relinquished the office of Prime Minister of Malaysia ('Malaysian PM Under Pressure to Quit' 2008). At that point, the admission that his "biggest mistake" was "to ignore cyber-campaigning on the Internet which was seized by the opposition" (Abdullah: Big Mistake to Ignore Cyber-Campaign' 2008), probably cemented in the minds of many the crucial place of the Internet in Malaysian politics and imaginary. Badawi's departure led to then Deputy Prime Minister, Najib Razak, a member of the elite with impeccable political pedigree, taking up the post of Prime Minister.

HYBRIDISING

Borrowing from the West while in the East is a practice that most govern-ments in Asia are familiar, partly because of the West's economic and military dominance in modern history and partly because of the legacies of colonisation. For the heads of nations in Asia, there is no stigma in such borrowing so long as these adoptions achieve the desired objective. The pragmatism of this attitude is encapsulated in Deng Xiaoping's procla-mation that it matters not whether a cat is black or white, so long as it catches the mice—不管黑猫白猫,捉到老鼠就是好猫—to explain his atti-tude toward Chinese economic policy (李彦增 2009), Mahathir adopted the same pragmatism, borrowing from the utopian rhetoric of early Internet and, adjoining it to the need for a new economic policy and direction to insert a vision of the future in the Malaysian imaginary where the knowledge economy would reign. As Mahathir himself declared (1995, p. 20):

> The Malaysian Government is, therefore, right in being pragmatic. Its acceptance of the capitalist free-market system is not total. It is condi-tional; an adaptation of the system to suit local conditions... The absence of rigid ideological tenets frees the Government to do what is practical and beneficial rather than what is ideologically proper.

Borrowing from the Wild Wild West of the Californian Ideology and applying a pragmatic attitude, the MSC was right from the start spoken of as "the careful creation of a region with an environment especially crafted to meet the needs of leading edge companies seeking to reap the rewards of the Information Age in Asia…We are taking a single-minded approach to developing the country using the new tools offered by the Information Age (Mahathir 1996/2002, p. 184). It was this very single-minded business-friendly attitude garnished with a dash of an utopian imaginary that led to an early framework of Internet governance in Malaysia aimed at facilitating commerce and technology transfer. Having learned that a transfer of values was less fluid than imagined during the Look East Policy, the Mahathir administration sought with the MSC mega project to harness the exuberance of the early Internet to carry Malaysians along but direct collective energies into the K-economy (knowledge economy) using a combination of policies, funding, concessions and incentives. Ironically, a transfer of values did occur but not the specific set that was articulated in envisioning of the MSC. While the architects of the MSC had foreseen change in the way Malaysians "live and work within the MSC" was envisioned (Mahathir 1996/2002, p. 184), it was the Internet's potential as a space for expression that Malaysians found of most utility.

To be sure, governance of the Internet in Malaysia from its earliest days to 2008 is inseparable from the history of its use and users. This meant that the political upheavals that occurred where the Internet was put to use as a conveyor of political messages and exchanges have to be taken into any account of Internet governance in Malaysia. The Internet as governed today is a consequence of the myriad innovations, choices and usage Malaysians have put it to. Internet governance in Malaysia is also a product of the matrix of social, cultural, economic and political histories, circumstances and changes that eventuated over time. A different set of users, a different socio-political order and unforeseeable developments, no matter how similar, would have produced a different model of Internet governance. For example, Indonesia also underwent its own *Reformasi* in 1998 where the Internet's affordances played a key role but its model of governance is nowhere similar to Malaysia's (Lim 2002). In the next chapter we will see how and why Internet governance in Singapore unfolded in the way it did.

Local contexts are important in how we go about understanding what is imagined and expected of the Internet but global and broader contexts are also vital in the consideration of how internet governance continues to change. The metamorphosis of the computer into a personalised interface that Apple's two Steves realised was a turning point in computing because it took the computer out of the lab and into homes and offices. Exactly how important a development that was to enabling broader internet access is only appreciable in hindsight, comparable perhaps only to the effect produced by the development of the first smartphone in 1994 by IBM (Smith 2018). Both of these were essentially engineering developments without which the embedment of digital media in the social, cultural, economic and political aspects of daily life today may well have not developed. This goes some way toward explaining the contemporary infrastructural turn in internet governance studies across the world, focussing on a range of hard- and software boundary objects such as internet daemons (McKelvey 2018) and access points (Lim 2018) to platform power (Gillespie 2018). Yet, we do well to understand that the invention of these technologies were as crucial as their availability to the average Malaysian. The former without the latter would have left these technologies on the high self where virtual reality headsets stayed because they were exotic but unaffordable gadgets during this same period. It is, therefore, important point to note that it is precisely because of the internet's ubiquity in daily lives that the governance of the internet is a pressing matter for all. Malaysia is no exception.

There are also, as this chapter demonstrates, factors that can strongly influence how, even seemingly global technologies, are introduced to a nation's public and its reception. The first Mahathir regime's perspective of the Internet as the harbinger of the knowledge economy had the effect of framing the Internet largely as a convergence of multimedia technologies that would stimulate economic growth. Seen through such a one-sided economic optic meant insufficient attention was given to the changes it could introduce to the Malaysian imaginary. That might have been the end of matters if not for the shock of *Reformasi*, which sufficiently motivated disgruntled Malaysians to take some charge of and responsibility for the socio-political discourse of the country and reclaim it from the hegemony of the ruling coalition and mainstream media. Singapore, as will be discussed in Chapter 3, took a very different approach to the Internet and its governance, viewing it as a suite of media and communication technologies right from the start. In the 2020s, poised on the cusp of establishing 5G networking, smart city traffic management (and surveillance) and e-payment systems, within the context of a US-China struggle for digital supremacy, there are urgent questions facing Malaysia and its model of Internet governance that we will discuss in Chapter 5.

BIBLIOGRAPHY

Abdulai, D. (2004). *Can Malaysia Transit into the K-Economy?* Subang Jaya: Pelanduk.
Abdullah: Big Mistake to Ignore Cyber-Campaign. (2008, March 25). http://www.malaysiakini.com/news/80354. Accessed 10 July 2020.
Amazon Startup Story—Fundable. (2020). https://www.fundable.com/learn/startup-stories/amazon. Accessed 1 March 2020.
Antons, C. (2006). Intellectual Property Law in Southeast Asia: Recent Legislative and Institutional Developments. *Journal of Information, Law & Technology, 1*, 1–10.
Barlow, J. P. (1996). *A Declaration of the Independence of Cyberspace.* https://www.eff.org/fr/cyberspace-independence. Accessed 2 March 2020.
Bill to Ensure Best Deal for Investors. (1996, August 2). *Business Times.*
Brand, S. (1998). *The Media Lab.* New York: Penguin.
Dreier, P. (2011, February 4). *Reagan's Real Legacy.* https://www.thenation.com/article/reagans-real-legacy/. Accessed 2 December 2019.
Funston, J. (2000). Malaysia's Tenth Elections: Status Quo, Reformasi or Islamization? *Contemporary Southeast Asia, 22*(1), 25–26.
Furuoka, F. (2007). Malaysia-Japan Relations Under the Mahathir Administration: Case Studies of the "Look East" Policy and Japanese Investment in Malaysia. *Asian Survey, 47*(3), 505–519.
Gillespie, T. (2018). *Custodians of the Internet: Platforms, Content Moderation and the Hidden Decisions That Shape Social Media.* New Haven: Yale University Press.
Hopkins, J. (2014). Cybertroopers and Tea Parties: Government Use of the Internet in Malaysia. *Asian Journal of Communication, 24*(1), 5–24. https://doi.org/10.1080/01292986.2013.851721.
ICANN History Project: ICANN's Early Days. https://www.icann.org/en/history/early-days. Accessed 2 July 2020.
Jomo, K. S. (1988). *Mahathir's Economic Policies.* Kuala Lumpur: Insan.
Khoo, B. T. (2003). *Beyond Mahathir Malaysian Politics and Its Discontents.* London and New York: Zed.
Leong, P. P. Y. (2019). *Malaysian Politics in the New Media Age Implications on the Political Communication Process.* Singapore: Springer.
Leong, S. (2014). *New Media and the Nation in Malaysia: Malaysianet.* London: Routledge.

Leong, S. (2016a). A Right and Not a Privilege: Freedom of Expression and New Media in Malaysia. In Larissa Hjorth & Olivia Khoo (Eds.), *The Routledge Handbook of New Media in Asia* (pp. 155–164). London: Routledge.

Leong, S. (2016b). "Is Allah Just for Muslims"?: Religion, Indigenisation and Boundaries in Malaysia. In C. Gomes (Ed.), *The Asia Pacific in the Age of Transnational Mobility: The Search for Community and Identity on and Through Social Media* (pp. 119–141). New York: Anthem Press.

Leong, S. (2009). The Hindraf saga: media and citizenship in Malaysia. In T. Flew (Ed.), *Communication, Creativity and Global Citizenship: Refereed Proceedings of the Australian and New Zealand Communications Association Annual Conference*. Kelvin Grove: QUT. https://eprints.qut.edu.au/34474/. Accessed 10 July 2020.

Lim, M. (2002, December). Cyber-Civic Space in Indonesia: From Panopticon to Pandemonium? *International Development Planning Review, 24*(4), 383–400.

Lim, M. (2018). Dis/Connection: The Co-Evolution of Sociocultural and Material Infrastructures of the Internet in Indonesia. *Indonesia, 105*, 155–172. http://doi.org/10.1353/ind.2018.0006.

Mahathir, M. (2002a). Forces That Will Shape Our Common Digital Future. In M. B. Mohamad, D. N. Abdulai, T. C. Ng, & N. T. Chuan (Eds.), *Mahathir Mohamad: A Visionary & His Vision of Malaysia's K-Economy* (pp. 105–118). Subang Jaya: Pelanduk.

Mahathir, M. (2002b). Inventing Our Common Future. In M. B. Mohamad, D. N. Abdulai, T. C. Ng, & N. T. Chuan (Eds.), *Mahathir Mohamad: A Visionary & His Vision of Malaysia's K-Economy* (pp. 153–162). Subang Jaya: Pelanduk.

Mahathir, M. (2002c). MIMO's Role in Malaysia's Move Toward a Knowledge Economy. In M. B. Mohamad, D. N. Abdulai, T. C. Ng, & N. T. Chuan (Eds.), *Mahathir Mohamad: A Visionary & His Vision of Malaysia's K-Economy* (pp. pp. 87–96). Subang Jaya: Pelanduk.

Mahathir, M. (2002d). The Digital Economy and the Borderless World. In M. B. Mohamad, D. N. Abdulai, T. C. Ng, & N. T. Chuan (Eds.), *Mahathir Mohamad: A Visionary & His Vision of Malaysia's K-Economy* (pp. 233–240). Subang Jaya: Pelanduk.

Mahathir, M. (1991). *Vision 2020: The Way Forward*. https://www.pmo.gov.my/vision-2020/the-way-forward/. Accessed 17 April 2019.

Mahathir, M. (1995). *The Malaysian System of Government*. Kuala Lumpur: Prime Minister's Office.

Mahathir, M. (1996). *The Opening of Multimedia Asia on Multimedia Super Corridor*. http://www.mahathir.com/malaysia/speeches/1996/1996-08-01.php. Accessed 2 July 2020.

Mahathir, M. (1996/2002). The Multimedia Super Corridor: Realising a Vision for Malaysia. In M. B. Mohamad, D. N. Abdulai, T. C. Ng, & N. T.

Chuan (Eds.), *Mahathir Mohamad: A Visionary & His Vision of Malaysia's K-Economy*. Subang Jaya: Pelanduk.

Malaysian PM Under Pressure to Quit. (2008, April 1). *Al Jazeera*. https://www.aljazeera.com/news/asia-pacific/2008/04/2008525135836481466.html. Accessed 10 July 2020.

McKelvey, F. (2018). *Internet Daemons: Digital Communications Possessed.* Minneapolis: University of Minnesota Press.

MSC Malaysia Bill of Guarantees. Multimedia Development Corporation. http://www.mscstatus.com/bill-of-guarantee-incentive. Accessed 1 July 2020.

Noor, F. A. (2013). The Malaysian General Elections of 2013: The Last Attempt at Secular-Inclusive Nation-Building? *Journal of Current Southeast Asian Affairs., 32*(2), 89–104. https://doi.org/10.1177/186810341303200205.

'People-Oriented' IT Agenda Required. (1996, December 19). *New Straits Times*, pp. 18–22.

Police Raid Offices of Malaysiakini.com News Website. (2016). https://rsf.org/en/news/police-raid-offices-malaysiakinicom-news-website. Accessed 10 July 2020.

Prime Minister's Office. (2006). *The National Mission 2006–2020*. https://www.pmo.gov.my/national-mission-2006-2020/. Accessed 10 July 2020.

Rawlinson, N. (2017). *Apple Was 41 Years Old In April, Here's Some History*. https://www.macworld.co.uk/feature/apple/history-of-apple-steve-jobs-mac-3606104/. Accessed 3 December 2019.

Saravanamuttu, J. (1988). The Look East Policy and Japanese Economic Penetration in Malaysia. In S. K. Jomo (Ed.), *Mahathir's Economic Policies*. Petaling Jaya: Insan.

Shariffadeen, T. M. A. (1994). *Information Technology and Development Malaysia's Experience*. Kuala Lumpur: MIMOS.

Smith, R. (2018). *IBM Created the World's First Smartphone 25 Years Ago*. https://www.weforum.org/agenda/2018/03/remembering-first-smartphone-simon-ibm. Accessed 10 July 2020.

World Bank. (2019). *World Bank Country and Lending Groups*. https://datahelpdesk.worldbank.org/knowledgebase/articles/906519-world-bank-country-and-lending-groups. Accessed 12 June 2020.

李彦增. (2009). 邓小平同志"黑猫白猫论"背后的故事. http://cpc.people.com.cn/GB/85037/8530953.html. Accessed 10 July 2020.

CHAPTER 3

The Internet in Singapore: From 'Intelligent Island' to 'Smart Nation'

Abstract This is a chapter about the historical, political and ideological underpinning and the variegated impacts of internet development in Singapore. It traces the development of the Internet in Singapore from the 1990s, when mass public internet access first began, through the 'intelligent' era (1990s to 2000s) and into the digital phase (circa 2010 onwards), culminating in the present 'Smart Nation' era (from 2015). The chapter highlights the strategic single-mindedness of Singapore's journey to become one of the most technological and digitally networked societies in the world. Even before the advent of the Internet, Singapore had already envisioned a first world city-state with top-rate communication infrastructure, one that would be supported by a highly-educated and technologically-savvy workforce. We see in this chapter how Singapore's determined pursuit of—and successes in—'intelligent' and 'smart' Internet visions has set the scene for it to influence proceedings in global debates around aspects of internet governance, especially in relation to political and economic sovereignty.

Keywords Singapore · Intelligent Island · Intelligent Nation · Smart Nation · Test-bed · Global hub

S. Leong and T. Lee, *Global Internet Governance*,
https://doi.org/10.1007/978-981-15-9924-8_3

INTRODUCTION

One important test case for understanding the relationship of information and communication technologies and democratization is Singapore. This city-state stands out internationally in two regards. First, the country's leaders exert a level of social and political control that is unique among wealthy nations. And second, those same leaders are engaged in one of the most far-reaching attempts to infuse information technology in society and make their nation an "intelligent island". (Warschauer 2001, p. 305)

Even before the advent of the Internet, Singapore had already envisioned a first world city-state with top infrastructure and communications connectivity, one that would be supported by a highly-educated and technologically-savvy workforce. Although it was the second country in Southeast Asia, after Malaysia, to offer public internet access in July 1994, once decision was made to adopt and embrace the internet, diffusion and penetration of the new technology was swift and unmistakeable. Since then, Singapore has been relentless in its pursuit of infusing information technology (IT) and information and communication technologies (ICTs) in Singaporean society (as cited in Warschauer 2001), and in the process, making technological and internet history/ies.

In 1995, Singapore became the first country in the world to host a national website: the Singapore Infomap at http://www.sg, which is still operational to this day. By 1999, Singapore had become the first country in the world to have virtually all homes and businesses connected to an island-wide hybrid fibre-optic cable network, enabling Singapore to attain the status of 'Intelligent Island' (a term originally coined by the British Broadcasting Corporation (BBC) in 1990 in a televised programme featuring Singapore's bold information IT developments), but became a buzzword for the global hub that Singapore has become known for.

This chapter introduces us both to the specifics as well as the general elements of Singapore's development of the Internet. It takes the view that the development of the internet in Singapore is best understood by looking at the myriad of government policies, or masterplans, and in the roll-out of infrastructure and technological imperatives that have enabled Singapore to thrive in the digital age. As such, chapter unpacks some of the key Internet plans that, chronologically and collectively, inform our understanding of Internet developments in Singapore since the mid-1990s, and in doing so, prepare us go understand the nation's

Internet governance mentality. It would also explain how these strategies have forged a society that embraces the internet and most social, cultural and economic facets of digital technology use, to the extent that Singapore is almost instantly thought of in technological terms globally. As the opening quote to the chapter by Warschauer makes clear, the Singaporeans leaders have "engaged in one of the most far-reaching attempts to infuse information technology in society and make their nation an 'intelligent island'" (2001, p. 305).

This chapter introduces and explains the key policy initiatives that underpin Singapore's Internet journeys, including:

- A Vision of an Intelligent Island: IT2000 Report (1992), the first plan designed to create a nationwide information infrastructure.
- Formation of two new statutory boards: the Infocomm Development Authority of Singapore (IDA) and the Media Development Authority (MDA), formed in 1999 and 2003 respectively to implement strategies to spread and speed-up the use of computing technologies in everyday life.
- Intelligent Nation 2015 (iN2015) master plan, launched in June 2006, to integrate all aspects of info-communications into a single ultra-fast broadband platform.
- Insights into Singapore's "Smart Nation" initiative, launched in November 2014 (just as iN2015 was to reach its promised fulfilment), which is very much premised on Singapore's successful development of its 1990s to 2000s 'intelligent' infrastructure (DGO 2018).[1]

The journey from moving Singapore from being intelligent adopters of the internet to smart users of digital technologies has been an extremely structured and deliberate one. To do so requires massive economic investment, and strong social, cultural and political capita—or to put it cogently, the Internet in Singapore has been developed with absolute buy-in from its citizenry. Indeed, Singapore is one of very few nations where even 'non-users' of the Internet (including those who consider themselves somewhat 'techno-phobic', even if they do not use the term) would readily identify themselves as supporters of internet use.

[1] https://www.smartnation.sg/.

The ideological perspectives proffered in this chapter are critical to making sense of the Singaporean mentality (and governmentality) in relation to a range of internet issues such as public acceptance of government regulation and intervention, national sovereignty, public surveillance and other principles of governance (that will be further addressed in Chapter 5).

INTERNET DEVELOPMENT IN SINGAPORE: THE HISTORICAL, POLITICAL AND IDEOLOGICAL

One important advantage which we have which we must take full advantage of is to use technology extensively and systematically, particularly IT. Not just piecemeal, individual gadgets, individual programmes and systems – that we are already doing, and all sorts of devices and applications have technology and IT in them. I am sure just in this room if we add all our handphones together we will have terabytes of storage and gigabytes of processing power but we have to do this systematically, to make the most of the potential, to integrate all of the technology and possibilities into a coherent and comprehensive whole. This will make our economy more productive, our lives better, and our society more responsive to our people's needs and aspirations. (Prime Minister Lee Hsien Loong, Smart Nation launch, 24 November 2014)

Launched in late 2014, Singapore's "Smart Nation" initiative represents a large-scale whole-of-government effort to apply digital modes of information and communications technologies (ICTs) to turn Singapore into a model city exemplified by the ubiquity of deployment of technological tools, devices and expertise. As Ezra Ho explains, a smart city is characterised by "extensive and systematic incorporation of digital networked technologies across the urban landscape and population" (Ho 2017, p. 3102). One may conjure up a vision of a smart city with densely connected assemblages of sensors, devices, objects, people and infrastructure that is supported by a wealth of data and algorithmic functions that would optimise processes in every aspect of urban life (Ho 2017, p. 3102). Indeed, at its inception, the Smart Nation initiative focused on urban digitalisation in five key domains: transport, home and environment, business productivity, health and tech-enabled ageing, and public sector services (DGO 2018). These have since been expanded, particularly with growing emphases on artificial intelligence and machine-learning technologies (Woo 2018). While we will return to touching on Singapore's Smart Nation initiative later in this chapter, it is quite apparent from the outset that Internet discourse—of which Smart Nation has exemplified since late 2014—in Singapore is typically spoken of using corporate narratives that privilege economic productivity. Put simply, the overarching narrative in Singapore is that smart use of the Internet and accompanying digital technologies in Singapore will lead to better lives (as articulated by Prime Minister Lee Hsien Loong in the quote above). In the context of Singapore, while there are regular concerns about the speed at which Singaporeans, particularly the less technologically-inclined, are able to adapt to smart living, most of such policy narratives and developmental visons are rarely challenged or seriously critique. This section of the chapter explains how and why Singaporeans have both actively and tacitly, concomitantly, imbibed a pro-developmental mindset when dealing with technological development. It does so by detailing a conceptual account of how the Internet space in Singapore has developed from the advent of mass Internet access (in 1994), through the era of web 2.0 (2000s), and into the Smart Nation vision (since 2014).

Before we proceed any further in detailing the broader context of Singapore, it is important to state—or perhaps, re-state—from the outset that contemporary Singapore exists as a duality, or indeed, as "necessary contradictions" (Lee 2010, p. 17). As political scientist Kenneth Paul Tan notes most cogently:

> Singapore is simultaneously a small postcolonial multicultural nation-state and a cosmopolitan global city of the top rank. This duality has produced a dynamic variety of contradictions and tensions, which have made all the more challenging the task of keeping the authoritarian system stable, durable, and successful. For one thing, the state has had to exert continuous effort to contain, rather than resolve, these contradictions in a pragmatic (that is to say, adaptive and undogmatic) fashion whilst directing the public narration of a coherent and persuasive story that is acceptably meaningful and even inspiring to Singaporeans and other countries looking for ideas, role models, and partners for development and governance. (Tan 2018, p. 2)

In the earlier years of Singapore's industrialisation under the leadership of its first and much-revered Prime Minister, the late Lee Kuan Yew, the

Singapore government found itself frequently refuting and defending its authoritarian style of rule. Singapore found itself dogged by labels such as 'police state' and 'nanny state', which it had to concurrently refute and tolerate as it sought to develop the nation into an economic powerhouse. This required a massive balancing act that involved nation-building and capacity-building internally, whilst utilising commercially motivated and outward-facing strategies to position Singapore as a cosmopolitan, safe and well-connected city of the future.

During this period of transition from the 1970s to 2000s, Singapore became well known—or perhaps, notorious—for being a politically censorious and highly-regulated society. With toilet-flushing and anti-spitting rules, as well as widely derided laws banning the sale and distribution of chewing gum, it is not too difficult to understand why Singapore has come under frequent insults and criticisms by those hailing from liberal democratic traditions. Indeed, Singaporeans are often described as living in a socially-engineered and 'self-censorious' climate of fear (Gomez 2000; Lee 2010). But at the same time, Singapore has defied most critics by being able to derive both economic and technological successes despite its rule-laden culture and supposed socio-political rigidity. With first-world infrastructure and a highly-educated and technologically-savvy workforce, Singapore is widely acknowledged as one of the most digitally networked societies in the world. Once Singapore hooked up to the World Wide Web back in 1994, Internet diffusion and universal access became swift and unmistakeable (Ang 2005; Lee 2010). While it is easy to dismiss Singapore's technological success as a result of its small geographical size—of about 700 square kilometres in total—one needs to bear in mind the fact that the city-state has been able to achieve rapid internet development whilst adhering to its model of authoritarian control and media governmentality (Lee 2010).

The economic and technological success of Singapore has meant that while the government still does not like being branded authoritarian by detractors—preferring instead to trumpet its many accolades, especially in the technological domain—it is less affected by negative portrayals in recent years. Not least because the proportion of global admirers of the Singapore narrative—which has also been referred to in some quarters as the 'Singapore model' (Ortmann and Thompson 2016)—has increased substantially in view of Singapore's indisputable economic success story. This success story is also mirrored in Singapore's Internet development, starting with the 'Intelligent' era first encapsulated in the 'Intelligent Island' and 'Intelligent Nation' discourses, which is where we turn our attention to next.

THE 'INTELLIGENT' ERA: 1990S TO 2010S

Singapore was an early and vigorous adopter of both IT and ICTs. IT was seen through the lenses of computerization for improved economic productivity in the early 1990s, which was in turn viewed a means of transcending the country's small human resource pool. For this reason, a National Computerization Masterplan was formulated in 1981. This saw the establishment of the National Computer Board (NCB) so that it would lead in the computerization of major functions in all government departments. Following this, a National IT Plan was established in 1986 to facilitate the development of an export-oriented IT industry and for raising productivity through the use of IT (George 2006). IT was used during this period mainly as an exercise in business efficiency, especially as Singapore saw its future in enhancing its port operations and boosting its trade-related services.

The 'Intelligent Island' nomenclature made its way into Internet governance folklore when the NCB produced what was to become Singapore's successor to the 1981 and 1986 IT masterplans. Unveiled in 1992, *A Vision of an Intelligent Island: IT2000 Report* became Singapore's flagship IT masterplan (NCB 1992; Lee and Birch 2000). The *IT2000 Report* or Masterplan envisioned a fully-wired nationwide information infrastructure by the year 2000, with every home and business connected via cable, and with IT permeating every aspect of society. This was a massive undertaking for a country that was still newly developing at the time. By the end of 1999, the IT2000 Report was successfully implemented not just in material or physical terms, but more significantly in ideological terms as the majority of Singaporeans bought into the view that rapid adoption and mastery of IT were necessary for the country to achieve sustained economic growth in the twenty-first century (Lee 2010, pp. 107–108).

The completion of the IT2000 Masterplan was a significant milestone for Singapore as it could deliver multi-channel television services via the predominantly government-owned Starhub CableVision. It also gave Singaporeans early glimpses into what future broadband services might look like, although at that stage, the cable network known as Singapore ONE (One Network for Everyone) offered little more than faster Internet and email access. Nevertheless, it gave credence to the Singapore government's widely proclaimed aim (first articulated in its 1986 National IT plan) to turn the country into an intelligent IT and digital media hub so that it can be transformed into an intelligent and creative knowledge-based economy. The phrase 'knowledge-based econ-omy' entered the national lexicon, to the extent that it was echoed in ministerial speeches and government reports through much of the 1990s. As an aside, it was observed then that most Singaporeans were really equating 'knowledge-based' with IT competence, even though they do not refer to the same knowledge base. IT2000 was nonetheless successfully rolled out, and subsequently succeeded by the Infocomm21 Masterplan, which had the aim of developing Singapore into one of the world's top five knowledge-based Information Societies (George 2006, p. 66).

By the mid-1990s, although wary about the ideological threats that accompanied democratic values that seemed inherent in the free flow of information, government leaders were consumed by the potential of the Internet in rapidly linking Singapore up with the world. Like its Malaysian counterpart, the Singapore government treated the Internet as an economic infrastructure from its very inception, carefully distin-guishing it from its political and libertarian 'origins' (Resnick 1998). This was not incidental but very deliberate. The government was particularly wary of a certain libertarian ideology that circulated around the advent of the global Internet at the time. It seemed tacit that the new Internet code meant that the individual would have the right, freedom or autonomy to choose and participate in political debates in a democratic fashion as illiberalism and authoritarianism would soon be defeated (Lee 2010, p. 104).

But Singapore had other ideas as it sought to not just understand, but master, the technology from a governance perspective. In other words, Singapore began to work through—almost with clinical precision—how it could maximise the business and commercial benefits whilst restricting access to what it perceived as 'undesirable material' in the same way it has done with traditional print and broadcast media in the offline world. Singapore's internet governance approach will be discussed at length in Chapter 5, but suffice to point out at this juncture that Singapore had begun to normalise the functions of the Internet by making it part and parcel of everyday life and ordinary politics very early on (Lee 2010, p. 105). Much the same was happening globally with many business entities and corporations starting to acquire Internet domain names and setting up their own websites, homepages and displaying their online interfaces. As Shapiro had already posited at the time:

We should not be surprised to see governments and corporations trying to shape the code of the [Internet] to preserve their authority or profitability. But code is not everything. Even if we lock in the democratic features of the Internet, the ultimate political impact of the Internet must be judged on more than design. We must also consider the way a technology is used and the social environment in which it is deployed. (Shapiro 1999, p. 15)

The Singapore government worked out very quickly that however global the Internet was poised to become, it existed nonetheless within the present sphere(s) of culture, economics and politics, not outside of it/them. As many internet studies researchers came to recognise, for most people, their Internet sojourns mostly traversed around their local communities, often writ large into seemingly global community/ies (Rheingold 1993). The strategy was therefore to continue to make it local by ensuring that the control point remains local, or it had to be based in Singapore socially, culturally, economically, technologically and, indeed, politically.

To continue to wield power over the users of the internet, the Singapore government worked out that it had to be involved in shaping both the socio-cultural and technological designs as well as regulatory environment in which the internet and other digital technologies operate (Lee 2010, p. 105). This marked one of the earliest articulations of the belief and value placed on geophysical Internet sovereignty globally, although it was not identified as such at the time. The local-global nexus that the Internet was—and still is—meant that the Internet had to be administered, controlled and regulated to 'order chaos' (Ang 2005). As Singapore honed its regulatory and Internet governance framework, it also set in place ambitious technical infrastructure development and pursued it with a high degree of precision, and obsession.

By 2015, as part of its latest *Intelligent Nation 2015* (iN2015) master plan launched in June 2006, Singapore had planned to integrate all aspects of info-communications (or infocomm for short, a commonly used term in Singapore) into a single ultra-fast broadband platform that would be capable of delivering internet access speeds of up to 1Gbps (IDA 2006), which was at the time seen as 'super-fast'. In a typical Singaporean style of wanting to be ahead of the developmental curve, this network was not only slated to be one of the best in the world, the accompanying social, cultural and political targets are just as ambitious, and most telling. According to the Infocomm Development Authority (IDA), the government agency responsible for regulating the info-comms industry and rolling out the iN2015 master plan (IDA 2006), Singapore was to become No. 1 in the world to harness infocomm services to add value to the economy and society by 2015. Whether or not this goal was achieved is not as important as the way in which the IT2000 plan was carried out through the 1990s, and indeed the obsessive manner in which Singapore had gone about embracing IT and attaining Internet supremacy in Southeast-Asia, and to a large extent, in the wider Asian region.

There was also the psychological aspect. On the surface, iN2015 was really just an upgraded IT2000 plan catering for the digital demands of the twenty-first century, not unlike the way personal computers, mobile phones and the software they utilise are upgraded every few years. There are, however, differences in the way terms like 'technology' and 'intelligence' are presented as critical human enablers—hence the shift from 'intelligent *island*' to 'intelligent *nation*' in the naming of the iN2015 masterplan, where the idea of the 'nation' (N) is deliberately emphasised. By this time, the government had decided that it was psychologically disadvantageous for Singaporeans to think of the country as an intelligent 'island' as it gave the impression of being marooned and therefore limited and constrained. Integrating an intelligent 'nation' to the digital world of the internet however would widen the scope of possibilities that infocomm and digital technologies could bring for a geographically-small city-state like Singapore (Lee 2010, p. 108). As the iN2015 Report noted quite specifically:

> [iN2015] recognises that infocomm alone will not be sufficient to transform the country's economic sectors. Neither will infocomm on its own change mindsets on how integration can yield benefits, how new opportunities can be realised by accessing international markets, or the extent to which service quality can be raised in an industry. However, infocomm can be a critical enabler to achieving all of these. Apart from boosting Singapore's economic competitiveness, infocomm will be used to enrich the lives of every individual. (IDA 2006, pp. 8–9)

Even prior to the unveiling of the iN2015 masterplan, the Singapore government's pro-technology mentality and general success in promoting high computer and Internet use had been consistently reflected in most statistical data. In 2002, a Singapore Internet Project (SIP) team, comprising Singapore-based scholars, thinkers and academics, published a book that offered a first-ever public report entitled *Internet in Singapore: A Study on Usage and Impact* (Kuo et al. 2002). The now-defunct SIP was part of the World Internet Project (WIP) network coordinated by research teams from Singapore's Nanyang Technological University (NTU) and the Centre for Communications Policy, University of California, Los Angeles (UCLA), collectively becoming the founding members of the WIP. The SIP was originally funded by the two government authorities responsible for managing the Internet's development during the first decade of public Internet access and services: the Singapore Broadcasting Authority (SBA) and the aforementioned Infocomm Development Authority (IDA) at the time. Subsequent studies were not as readily available in the public domain as government authorities in Singapore took over the role of tracking data, and other statistical analyses.

The SIP report presented the 1999 findings from what became the benchmark survey (that was to have led to a longitudinal study in 2000 and beyond) that monitors Internet usage and its social impact. The newsworthy aspect of the report was that by 1999, Singapore's status as a technological society with high IT penetration and use was already high by global standards, with nearly 50% of adults, age 18 and up, were active users of the Internet. The numbers were markedly higher for students, with internet penetration at 71% (and rising) (Kuo et al. 2002, p. 100). Peculiarly for Singapore, even 'non-users'—defined in the report as people who do not access the Internet due to three key reasons: did not know how, no time and no interest—were found to be generally supportive of the Internet.

Andrew Barry, in his landmark study of how a technological society can be governed, alluded to the importance of having citizens on side with governments who desire to succeed in the technological age that was dawning back in the 1990s and early 2000s (Barry 2001). As Barry points out, while citizens of a technological society are "expected to have a certain knowledge of technology, and to make choices on the basis of this knowledge", not everyone will be "willing or able to meet these expectations" (Barry 2001, p. 29). He goes on to contend that at the very minimum, it is crucial that such people do not become hindrances to the process of turning the internet and ICTs into essential tools and technologies for the conduct of business and everyday life. While Barry does not specifically highlight Singapore in his work, Singapore fits this criterion beautifully. Since the start of 2000, with the fulfillment of the

IT2000 network, Singaporeans have indeed acquired the capacity to go digital in most aspects of their everyday lives—if they wish to. Unwillingness or disinterest in the Internet, especially among the economically productive generation, became less of a social or statistical concern in Singapore, so much so that the Singapore government has come to define the digital divide as "the gap between those who are internet savvy and those who are not" (George 2003; cited in Lee 2010, p. 109). This gap has since been drastically reduced with Singapore boasting individual internet penetration rate of 87% and mobile penetration rate of nearly 150% as at 2018.[2]

This section tells us that Singapore's pro-technology mentality was already in full force before the what we refer to next as the 'digital era', led by web 2.0 technologies, took off. In fact, since 2001, Singapore has already become more digitally connected than developed countries like the United States of America, Australia and the United Kingdom (Lee 2005, 2010). Although innovative uses of the Internet continue to emanate from Silicon Valley and other tech centres of the world (such as Japan, India, South Korea, Taiwan, and later on, China), Singapore was ready to embrace the digital era through its 'Smart Nation' policy and embark on plans to capitalise on the growth of big data, artificial intelligence, machine-learning and other forms of digital and algorithmic applications.

The Digital Era: in2015 to 'Smart Nation'

By the time 2015 arrived, the Internet had already entered what was popularly known at the time as 'Web 2.0' status. This was marked by the intensification of digital convergence, the advent of mobile and smart telephony and formation of social media networks. The three principal new government agencies that were formed in Singapore through the 2010s to consolidate regulatory management of the Internet, digital and IT services include:

- Government Technology Agency (GovTech), formed in 2016 following a restructure of the former IDA. Housed directly under

[2] Individual Internet access rate: https://data.gov.sg/dataset/individual-internet-access; mobile telephony penetration rate: https://data.gov.sg/dataset/mobile-penetration-rate.

the Prime Minister's Office, GovTech is responsible for the delivery of the Singapore government's digital services to the public. It is the agency that provides the infrastructure to support the implementation of Singapore's Smart Nation initiative designed to harness IT, digital networks and big data. New offices and departments were subsequently set-up (such as the Digital Government Office, or DGO, to lend greater support to the Smart Nation vision).

- Infocomm Media Development Authority (iMDA), formed in 2016 through the restructuring and merger of the Info-communications Development Authority (IDA) and the Media Development Authority (MDA). iMDA undertakes the role of regulating the converged information, Internet and media sectors in Singapore, but it is also tasked with ensuring data protection in Singapore through the Personal Data Protection Commission.
- Cyber Security Agency of Singapore (CSA), formed in 2015, is a government agency, under the Prime Minister's Office but administratively managed by the Ministry of Communications and Information, that provides centralised oversight of national cyber security functions, and works with the public and private sectors to protect Singapore's critical services and sectors.

While Singapore often receives accolades for its technical infrastructure, its ability to control the ever-expanding Internet space is not often understood or discussed. It is important to recognise that Internet governance in Singapore, while strongly led by public servants and experts working in one or more of the abovementioned government statutory bodies, is typically centrally managed and coordinated. By way of example, one of the earliest impactful decision by the CSA, back in August 2016, was to direct all government bodies and agencies to cut off web access for public servants as a defence against potential cyber attack (Wagstaff and Aravindan 2016). This approach took the public sector by surprise as it seemed counter-intuitive to most onlookers as it looked like Singapore was not only backtracking on its 'Smart Nation' initiative, but also returning to its pre-Internet past.

While critics labelled this move as a "a retreat for a technologically advanced city-state that has trademarked the term 'smart nation'", the decision proved to be a smart move in its own right as it did not cause major inconveniences. Not only was the government aware that most, if not all, civil servants already had mobile Internet access via their own digital devices at the time, it was also confident that the networked mentality had already been deeply ingrained into the psyche of all Internet users. The government was seen to prioritise security ahead of convenience, leading to greater trust among citizens and users of e-government services (Wagstaff and Aravindan 2016). In any case, as long as infrastructure wifi access remained universally available—which is the case in most public places in Singapore—public servants would not feel digitally disconnected.

Without taking an ideological vantage point or placing a moral judgement call, the history of Internet development in Singapore demonstrates that its success in stems from its ability to enhance favourable uptake of technology and Internet use whilst enabling commercial imperatives and technological innovation to flourish (Lee, 2010). Singapore adopts a more instrumental view of technology that differs from either a liberal or authoritarian/totalitarian approach (Feenberg 1991). Whilst the former subscribes to a democratic view of technology as being inherently free, and therefore liberal, the latter believes that information must be subject to regulatory forces and strict controls. In other words, the Singapore government takes a somewhat middle ground, believing that the Internet and all new technologies can not only be tamed and controlled, they can even be used to increase and strengthen centralized surveillance and control (Warschauer 2001, p. 305).

As demonstrated in its decision to exclude web access to its own civil servants from August 2016, when, how and under what circumstances Internet controls are carried out makes all the difference between extensive, tight-fisted controls and looser modes of supervision. In Chapter Five, the discourse of 'auto-regulation', understood in the main as automatic modes of control premised on the French philosopher Michel Foucault's (1977) discourse of disciplinary power will be unpacked (see Lee 2010). More specifically, we explain how the Internet in Singapore is being strategically regulated and policed, with a careful eye on not upsetting its economic potential, via the discourse of auto-regulation that has been enacted, even perfected, in Singapore.

Digital Smarts

In June 2019, the Singapore government hosted an international Smart Nation Summit at the Sands Expo and Convention Centre at Marina Bay, the most iconic venue in contemporary Singapore's central business district. This was an event attended by 15,000 local and international delegates, and therefore the ideal site to flaunt Singapore's smart city achievements and ambitions (Tham 2019, p. 4). At the closing dialogue session of the Summit, Prime Minister Lee Hsien Loong delivered a brief speech at which he disclosed the reasons why Singapore has the ingredients to attain Smart City status:

> We have a strong base to become a smart nation because the population is technologically literate, and IT infrastructure is good, internet access is fast and affordable and we have very high usage rate. There is fibre to every home, and a mobile phone penetration rate about 150%. It is not quite the highest in the world, but it is higher than you really need just to get by. Also, the government takes the lead. We were one of the first governments in the world to computerise, to digitise our data, and to move services online. We have services which are efficient, convenient and popular with citizens. Whether it is applying for a passport, paying taxes, paying your parking fines, it does not take more than a moment because the QR code is provided to you. (Lee, 26 June 2019)

The mainstream media summarised the Prime Minister's speech quite accurately by declaring that there are essentially three ingredients for success in Singapore's Smart Nation vision, namely: strengths in IT and engineering capabilities, a government with leaders who understand technology and a society that does not fear science or technology (Tham 2019, p. 4; Baharudin 2019, p. 1). Understood in this way, Singapore's Smart Nation initiative looks certain to be achieved in the near to medium term future, not least because it has been in the works prior the era of mass internet use. As Woo has pointed out, it is important to note that "the Smart Nation initiative was not established tabula rasa, but rather, represents a culmination of earlier efforts" at e-government programs or public service delivery (Woo 2018, p. 2). This aspect of the Smart Nation initiative was concretised when a new Digital Government Blueprint was unveiled in June 2018, which had at its core a promise for all Singaporeans to access up to 95% of government services online by 2023. This would involve every person having a digital identity, and for 20,000 public servants to be trained in data science in order to meet this goal. Singapore, according to the blueprint would be "digital to the core" (Tham 2018). It is palpable that the aim of Smart Nation is to make the entire delivery of government services not just more efficient, but also far more comprehensive in that it would involve the private sector as well as all commercial operations that are people-centric, including health, transport, urban solutions, finance and education. Even the ubiquitous hawker food stalls have not been excluded as many of them have been enticed to accept digital payment options since 2019 (Wong and Heng 2019).

In framing its Smart Nation vision, the government has expressed its desire to emulate the success of the three big Chinese tech companies—Baidu, Alibaba and Tencent—in facilitating digital financial and banking transactions not just within China, but increasingly globally (Lee 2019). But it wants to do more than just emulate China; nor does it hope to catch-up with China's digital economy, not least because it would be a futile (and foolish) exercise given the vast difference in the size of China's population, size and financial might. Singapore's long-held ambitions to be a hub economy of Asia in as many sectors as it can reasonably grow and develop has always been premised on being a 'test-bed' for new ideas, inventions and industries (Lee 2016). Indeed, it's overarching aim is to continue to wield economic influence as "the world's foremost living lab for widespread deployment of new technologies from fibre-optic internet to online government services", and much more (Khana 2018).

The term 'test-bed' was much used in the Singapore public service through much of the 1980s and 1990s as the government sought to incentivise multinational firms from around the world to establish their research and development (R&D) operations in Singapore. Tax breaks, along with fiscal support in the form of grants and open access to talents, were often dangled as incentives. Singaporeans were persuaded, perhaps socially-engineered, to support the government's mission in the form of getting themselves trained, skilled, and educated in the right discipline areas, especially in engineering, or engineering-related fields through the 1980s to 1990s, and in IT and biological/life sciences from the 2000s. While it is not untrue that Singaporeans would reap the benefit in terms of technological and knowledge transfer over time, the broader population became inured to the idea of being 'guinea pigs' in a laboratory, as well as having to constantly stay up to date with all forms of social changes.

To keep up, Singapore has realised that it needs to be nimble and be a fast-adopter of new technologies across a wide spectrum of areas, including 5G mobile telephony and broadband, cashless transactions, mobile banking, fintech services, smart street lamps and other internet-based digital tools. It has already embarked on widespread equipping of people through education and training, and established the necessary government departments, agencies and research facilities. In essence,

Smart Nation as a policy tool tells us—amidst the material and key performance indicators—that Singapore will spare no efforts in reaching their stated goal. As the Smart Nation policy document states in explicit terms:

> Smart Nation is integral to Singapore's next phase of nation building. To continue to prosper and stay relevant in the world, Singapore needs to ride the waves of the digital revolution and capture the opportunities it brings, just as we embraced globalisation before. Digitalisation presents opportunities for Singapore to enhance Singapore's traditional strengths, address and overcome our national challenges and physical limits, be it resource constraints of an ageing population, as well as build new sources of comparative advantage for Singapore. This new era of digital transformation will power Singapore to SG100 and beyond. (DGO 2018, p. 5)

Conclusion

It appears at first glance that Singapore has had it relatively easy in its development of Internet-based infrastructure and economy. Yet the Smart Nation drive, its most ambitious vision to date, tells another story. On the one hand, Singapore has been able to link the Internet with its broader economic mission of becoming a global economic hub; on the other hand, it had to apply visionary lenses to predict where the future of technologies was headed towards. As a small country deeply reliant on the flow of trade and services, it had to take a calculated gamble on becoming at worse the middle-person in the fast-developing digital world, not unlike its shipping and aeronautical port facilities and services in the offline world; and at best, a leader in electronic commerce and digital services (Lee 2016). Either way, developing its Internet and digital infrastructure was necessary in order to even join as a player. Having succeeded in being a recognised player, it now seeks to take a further step into a leadership role. As a small country, it has to continue jostling for recognition and a seat in the decision-making fraternity of the global digital economy, one that is intertwined with global Internet governance.

While Malaysia sought to plug in via its Multimedia Super Corridor project, Singapore's scope was just a bit broader as it sought to capitalise on any digital projects that could contribute to its GDP, provide job opportunities and had the potential to strengthen Singapore's hub economy status and stretch its socio-economic influence towards East Asia and the rest of the world. Singapore's desire in the digital realm is consistent with its broader political remit, which is to punch above its weight globally. As we will see in Chapter 5, this applies as well to its global internet governance approaches.

Bibliography

Ang, P. H. (2005). *Ordering Chaos: Regulating the Internet*. Singapore: Thomson Learning.

Baharudin, H. (2019, June 27). New Office Aims to Create 10,000 Jobs in Tech over 3 Years. *The Straits Times*, p. 1.

Barry, A. (2001). *Political Machines: Governing a Technological Society*. London and New York: Athlone Press.

Digital Government Office (DGO). (2018). *Smart Nation: The Way Forward*. Singapore: Smart Nation and Digital Government Office.

Feenberg, A. (1991). *Critical Theory of Technology*. New York: Oxford University Press.

Foucault, M (1977). *Discipline and Punish: The Birth of the Prison* (A. Sheridan, Trans.). New York: Random House.

George, C. (2006). *Contentious Journalism and the Internet: Towards Democratic Discourse in Malaysia and Singapore*. Singapore: Singapore University Press.

Gomez, J. (2000). *Self-Censorship: Singapore's Shame*. Singapore: Think Centre.

Ho, E. (2017). Smart Subjects for a Smart Nation? Governing (Smart)mentalities in Singapore. *Urban Studies, 54*(13), 3101–3118.

Infocomm Development Authority (IDA). (2006, 20 June 20). *Singapore iN2015 Masterplan Offers a Digital Future for Everyone*. IDA Media Release. http://www.ida.gov.sg. Accessed 4 June 2010.

Khana, P. (2018, July 24). What Does a Smart Society Look Like? *The Straits Times*. www.straitstimes.com. Accessed 24 July 2018.

Kuo, E. C. Y., et al. (2002). *Internet in Singapore: A Study on Usage and Impact*. Singapore: Times Academic Press.

Lee, H. L. (2014, November 24). *Speech by Prime Minister Lee Hsien Loong at Launch of Smart Nation Initiative*. Singapore. https://www.smartnation.sg/whats-new/speeches/smart-nation-launch. Accessed 18 February 2020.

Lee, H. L. (2019, June 26). *Speech by Prime Minister Lee Hsien Loong at the Smart National Summit Week Closing Dialogue*. Singapore. https://www.pmo.gov.sg/Newsroom/PM-Lee-Hsien-Loong-Smart-Nation-Summit-Week-Closing-Dialogue. Accessed 23 March 2020.

Lee, T. (2005). Internet Control and Auto-regulation in Singapore. *Surveillance and Society, 3*(1), 74–95.

Lee, T. (2010). *The Media, Cultural Control and Government in Singapore*. London and New York: Routledge.

Lee, T. (2016). Forging an 'Asian' Media Fusion: Singapore as a 21st Century Media Hub. *Media International Australia, 158*(1), 80–89.

Lee, T., & Birch, D. (2000). Internet Regulation in Singapore: A Policy/ing Discourse. *Media International Australia Incorporating Culture and Policy, 95*, 147–169.

National Computer Board (NCB). (1992). *A Vision of an Intelligent Island: The IT2000 Report.* Singapore: Singapore National Printers.

Ortmann, S., & Thompson, M. R. (2016). China and the "Singapore Model". *Journal of Democracy, 27*(1), 29–48.

Resnick, D. (1998). Politics on the Internet: The Normalization of Cyberspace. In C. Toulouse & T. W. Luke (Eds.), *The Politics of Cyberspace: A New Political Science Reader* (pp. 48–68). New York: Routledge.

Rheingold, H. (1993). *The Virtual Community: Homesteading on the Electronic Frontier.* Irvine, CA: Ingram.

Shapiro, A. L. (1999). Think Again: The Internet. *Foreign Policy, 115*, 14–17.

Tan, K. P. (2018). *Singapore: Identity, Brand, Power.* Cambridge: Cambridge University Press.

Tham, I. (2018, June 6). Almost All Government Services in Singapore to Go Digital by 2023. *The Straits Times.* www.straitstimes.com. Accessed 8 June 2018.

Tham, Y. (2019, June 27). PM Cites Three Ingredients for Success in Smart Nation Vision. *The Straits Times,* p. 4.

Wagstaff, J., & Aravindan, A. (2016, August 24). Mind the Air-Gap: Singapore's Web Cut-Off Balances Security, Inconvenience. *Reuters.* https://www.reuters.com/article/us-singapore-internet/mind-the-air-gap-singapores-web-cut-off-balances-security-inconvenience-idUSKCN10Y2F1. Accessed 29 February 2020.

Warschauer, M. (2001). Singapore's Dilemma: Control Versus Autonomy in IT-Led Development'. *The Information Society, 17*, 305–311.

Wong, L., & Heng, M. (2019, June 29). Unified Cashless Payment System Rolled Out at 500 Hawker Stalls Across Singapore. *The Straits Times.* www.straitstimes.com. Accessed 23 March 2020.

Woo, J. J. (2018). *Technology and Governance in Singapore's Smart Nation Initiative.* Cambridge, MA: Ash Center for Democratic Governance and Innovation, Harvard Kennedy School.

CHAPTER 4

Internet Governance—The Malaysia Way

Abstract This is a chapter that examines the arena of internet regulation in Malaysia between the years 2008 and 2018, two decades following the promise not to censor the internet. It positions the issue of internet governance in a regional context where the internet's ubiquity in Southeast Asia and its users' greater sophistication have also led to its abuse as a medium of communication. In Malaysia, unchecked by a complicit fourth estate, internet regulation in the name of cybersecurity was used to broaden the scope of extant Acts to include digital media, enact new laws and curtail political satire and informal sharing of news. Malaysia is, for example, the first country within ASEAN (Association of Southeast Asian Nations) to enact an anti-fake news law, which has opened the door for other countries in the region to further tighten regulation of the internet. The influence of Malaysian reform of internet regulation on global internet governance is yet to be determined. However, when viewed from the major impact that digital media liberation has brought to the nation, it suggests there to be many lessons on internet governance that can be derived from the Malaysian experience.

Keywords Bersih · Anti-Fake News Act · DFTZ · Anti-Westernism · Internet sovereignty

Much of how governance of the Internet occurs in Malaysia is similar to how it occurs elsewhere in the world. Governments have generally found themselves acting retrospectively rather than with foresight when it comes to the Internet. This is hardly surprising considering how many ways the Internet can be used and misused. As the array of internet-related technologies became widely available, new laws have had to be enacted to facilitate their rollout into societies. In Malaysia, this included the Payment Systems Act 2003, the Electronic Commerce Act 2006, the Electronic Government Activities 2007 and the Personal Data Protection Act 2010. However, where Malaysia and, to some degree, Singapore are different is in the expedience with which they en/act the necessary laws. Being semi-authoritarian states with strongman leaders in the decades of nascent nation-building brings that 'advantage' to the fore. Viewed from a liberal perspective, these are lamentable states of being for a sovereign people but viewed from a pragmatist democracy, no more than expected. There are, of course, exceptions to such circumstances within Southeast Asia, namely Indonesia and the Philippines but each of these can also be said to have derived their own hybrid models of IG, each differently tinged by colonialism.

In Chapter 2, we saw how myriad factors, some unforeseeable, and some very obvious, such as leadership expectations of technological change, shape and feed into how technologies are introduced to and received in Malaysia. This chapter continues to trace the twists and turns between 2008 and the present that have led to the current model of IG in Malaysia. Again, we employ here a socio-historical approach that links politics, economics, social and cultural transitions to technological governance. However, it would be remiss and dismissive of the complexities of socio-technological interactions to make direct correlations between any one event/incident and changes in Internet law and governance. Nor, with a book of this length, is it impossible to detail each and very turn made, even if the many intertwined connections could be identified. What we point to here, instead, are accretions of major and minor inflections in the dozen years or so that push governments toward specific acts and decisions and, the current state of IG in Malaysia.

If a Crime Is Committed …

After the 2008 General Elections and in between the various Bersih rallies, the BN government began to shore up its own online media competencies in the contest for public approbation (Vee 2011). "Cybertroopers" were recruited to guard and advance the interests of the sitting government (Hopkins 2014). Upon taking up the reins of government Prime Minister Najib Razak (hereafter, Najib) launched a new initiative aimed at uniting the country, dubbed 1Malaysia in 2009. According to Noor, "the shift to 1 Malaysia was significant in the sense that this was the first time that Malaysia's national political arena was being reconfigured according to a nation-building narrative that did not bear any traces of religious communitarianism" (2013, p. 96). In pursuit of broad appeal, Najib even promised to repeal the much-hated Internal Security Act (ISA). However, its replacement with an equally stringent law, the Security Offences (Special Measures) Act 2012 or SOSMA did little to convince Malaysians of the merits of the 1Malaysia project. Ultimately, Najib's inclusive and non-sectarian vision of the nation and the out-and-out wooing of non-Malay voters the 1Malaysia project, failed to gain much support (Noor 2013).

In contrast, Bersih's message of urgent electoral reform, built around understandings of constitutional law and human rights struck a chord with many in Malaysia. Perhaps this was because the Bersih coalition also began multiple campaigns (both online and offline) of public education on the rights of citizens, the power of votes and why, as well as how, they should exercise their ballot to greater effect in national politics. Or perhaps Malaysians were, by 2009, highly sceptical of any government initiatives and rhetoric about unity, nation-building and progress. Overall, in the years between 2009 and 2013, another four Bersih rallies were organised, with a second occurring in 2011, third in 2012, a fourth in 2015 and the last, in 2016.[1] Significantly, from the second Bersih rally onwards, the Malaysian diaspora become involve in homeland politics. Their involvement brought some financial support, foreign-trained talent and more importantly, drew international attention to the need for free and fair elections in Malaysia. With each rally thousands marched, across the country and the globe. Yellow became the colour of lawful dissent.

[1] http://www.bersih.org/.

In response, the Najib administration ordered that the riot police and water cannons be set on protestors and, fired tear gas into the crowds. A multitude of individuals were arrested, in one case for "behaving in an insulting manner in public under the Minor Offences Act" by dropping yellow balloons on the Prime Minister at a public event (Koya 2018). The Internet was the medium through which many aspects of the protests were organised, from the education of citizens, engagement with the domestic and international community of Malaysians, mobilisation of dissenters, arrangements for legal advice and representation in case of arrest as well as shifting the protest groups among different loci during the rallies themselves. To be sure, the Internet did not create the protests but its facility of communication between disparate groups made organisation easier and smoother, at once broad and narrowcasting via instant messaging, the WWW and internet telephony software such as Skype and Zoom. By 2011 the exasperated Najib administration professed itself ready to "contravene the MSC's bill of guarantees" (Razak 2011). And so the guarantee that inadvertently created the space for disenfranchised Malaysians to raise questions, discuss issue and practise dissent was muted with the insertion of a caveat so that it now reads (MIDA 2019, p. 4, my emphasis):

BoG 7: To ensure no censorship of the Internet
Objective:
To realize the vision for Malaysia to be a major global ICT hub, the Government recognizes that the revolutionary role of the Internet in facilitating information-exchange and innovation, and providing the basis for continuing enhancements to quality of work and life

- While Government will not censor Internet, *this does not mean that any person may disseminate illegal content with impunity & without regard to the law. To the extent that any act is illegal in the physical world, it will similarly be outlawed in the online environment. Laws prohibiting dissemination of, e.g. indecent / obscene or other illegal materials will continue to apply*
- Relevant ministries & Agencies will take appropriate actions to enforce laws under their purview

In the contest between a beautifully crafted ideal of "universal" IG surrounding the borderless world of the knowledge economy and the needs of a government threatened by a vocal citizenry, political expediency won hands down. So much so that on 1 May, just a few days before the 13th General Election of Malaysia took place on 5 May 2013, the Chairperson of the Malaysia Communications and Multimedia Commission (MCMC), Mohamed Sharil Tarmizi, reiterated: "if a crime takes place, whether it is in the virtual or real world, it still constitutes an offence. There is no difference" (MCMC summons 15 over spreading of rumours 2013). The world became borderless in a vastly different sense from that envisioned by Mahathir in the establishment of the MSC. Rather than a free flowing exchange of knowledge, ideas and technological innovations facilitated by the Internet, what occurred was the curtailment of information exchange through regulations and governmental perspectives that transcend the boundaries across the real/virtual divide. On 5 May 2013, Malaysia had its 13th General Election of Malaysia. The election became very much a contest between the *Barisan Nasional* coalition led by Najib Razak and *Pakatan Rakyat* (People's Alliance), led by former DPM Anwar Ibrahim. BN stayed in power with 133 out of 222 seats but lost significant ground to the opposition, garnering just 47.38% of the vote.[2] This led to accusations of gerrymandering that saw outspoken activists such as Adam Adil and opposition figures such as Tian Chua and Haris Ibrahim arrested and sentenced under the Sedition Act and the Penal Code in 2013 (Lawyer Haris Ibrahim gets eight months' jail for sedition 2016). Seizing upon the electoral win as the mandate to act, the Najib administration went on to propose the Prevention of Terrorism Act 2015 (POTA) and the National Security Council (NSC) Bill 2015. Many of these laws were aimed, in some way or form, at curtailing the wave of protests that unfurled across the years. The Peaceful Assembly Act, for example, stipulated that only citizens of a minimum of 21 years of age have the right of assembly in Malaysia. Additionally, street protests were made illegal without with the express permission of the police based on advance written notice containing personal details of organisers and speakers (Whiting 2011). Most certainly untenable restrictions designed to daunt would-be protest organisers.

[2] Election Guide. 2013. http://www.electionguide.org/elections/id/544/, 5 May.

China's Courtship and Anti-Westernism Redux

Bearing in mind the need for the broader contexts inclusive of technological trends and global patterns in usage, it is as well to remember that 2007 was the year when the iPhone was launched and smartphone ownership very quickly became a marker of coolness. In Malaysia, mobile phone ownership shot up from 85.1% in 2007 to 106.2% in 2009 although only 16.3% of handphone (mobile phone) users accessed the internet through their phones (MCMC Annual Report 2009, pp. 28 and 29). At that point, high speed broadband was being explained to the populace even as it was slowly rolled out across the country. By 2013, cellphone ownership was at 143.8% (MCMC Annual Report 2013, p. 170), meaning that many Malaysians own more than one mobile phone at the same time. Surveys revealed that a year later, 71.4% of handphone users confessed to being obsessed by their handphones even though only 53.4% owned a smartphone (MCMC Handphone Users Survey 2014, pp. 23 and 29). Smartphones from China and elsewhere in the world are freely available in Malaysia today, all the brands from the higher end Huawei and middle band Xiaomi to Oppo can be found at retail outlets and online shopping sites. By 2018 it was reported that 78% of Malaysia's population are smartphone users and nearly 94.6% of those who access the Internet do so through their smartphones (MCMC Annual Report 2018, p. 8). The growth of mobile Internet use is a global industry and user trend that has implications for IG. How do we get users today who are able to access information so easily to question, let alone verify, the veracity of the information received? How do we get users to pause and think before they hit the 'share' button? Is media literacy alone sufficient to filter rumour from fact when disinformation is rife? Or should some form of governance be put in place to prevent the exponential spread of information without evaluation? Are governments such as India (Jash 2019) and Iran (Lunden 2019) right to think shutting down the Internet to stifle political unrest an acceptable measure of governance? If so, how would the global nature of the internet and, if we are honest, the global flow of capital and data be affected? How can the Internet retain its role as a global space for creating and sharing information freely without, even if inadvertently, exacerbating tensions in local issues and contests? There is no lack of questions but very few broadly acceptable answers. Is it possible to implement effective IG that is truly apolitical?

Take the example of the answer the Najib Administration came up with: the Anti-Fake News Act 2018, which was enacted in April and defined "any news, information, data and reports, which is or are wholly or partly false, whether in the form of features, visuals or audio recordings or in any other form capable of suggesting words or ideas" as fake news so long as the authorities deemed it so (Attorney-Governor's Chambers 2018). Additionally, the Act also held intermediaries liable and co-opted them into the policing of fake news with less than prompt action in removing fake news from circulation with immediate effect (Malaysia Anti-Fake News Act 2018). Like India, much of this Act was ostensibly aimed at curbing the spread of rumours and falsehoods online. However, the enactment of the Act itself was not entirely without ulterior motive. Pushed through in the last days before the 2018 general elections, the Act was also an attempt to control allegations of PM Najib's corruption in the affairs known as 1MDB (One Malaysia Development Fund). Uncovered by the online portal, *The Sarawak Report,* reports of Najib's mismanagement of the nation's sovereign fund and attempts to cover it up stretch all the way from February 2015 to September 2016. Also the subject of two books, *The Sarawak Report: the Inside Story of the 1MDB Exposé* (Rewcastle-Brown and Brown 2018) and *Billion Dollar Whale* (Hope and Wright 2018*)*, the scale of corruption revealed outraged Malaysians and was hugely embarrassing for the Najib administration. While there is no way to directly correlate the two, it is fair to say the 1MDB scandal played no small part in *Barisan Nasional* finally losing government after more than six decades as the ruling coalition. At the time of writing Najib Razak and his associates are still being tried for their alleged crimes. When the new Pakatan Harapan government came into power in May 2018 it very quickly fulfilled at least one election promise and repealed the Anti-Fake News Act (Yeung 2018).

All this while, China, the Asian behemoth began earnestly in 2013 to supplement its soft power campaign with a more concrete plan, the Belt and Road Initiative (BRI) to woo the countries of Southeast Asia. Malaysia welcomed Chinese investments and know-how warmly and in 2016, the government launched the National eCommerce Strategic RoadMap with a view to "double the growth of its eCommerce market" (Malaysia's National eCommerce Strategic Roadmap 2016). Jack Ma, then founder of China's wildly successful Alibaba, was appointed Digital Economy Advisor to the Malaysian government in 2017. Later that same year, Malaysia launched the Digital Free Trade Zone (DFTZ), the first overseas base of Alibaba's eWTP (Electronic World Trade Platform) (Leong 2018, p. 74). By the end of 2017, several highly significant infrastructure projects such as the East Coast Rail Link (ECRL), the Melaka Gateway port, the Robotic Future City in Johor and the Kuala Linggi port were China-backed projects or China-Malaysia joint ventures. Interestingly, Jack Ma's take on digital technologies is very much in line with Najib's technocratic approach, where technology is generally regarded as neutral, neither inherently good or bad but knowledge to be utilised by smart individuals for the betterment of humanity. At the launch of the DFTZ on 3 November 2017, Ma very briefly but pointedly derided the hand-wringing of other countries:

> We believe [in] the future. Any problem that we encounter we have to solve it. ... I hate some of the countries, you know like many of the European countries, they worry, you know, ahhh... privacy, security ... all these things. The more you worry about the future, you don't have the future.

In other words, when it comes to business with China there is no need to fret over issues of human rights, democracy and civil liberties. Or indeed, the embedded prejudices, biases and injustices embedded in technologies such as algorithms, artificial intelligence, facial recognition technologies, all of which are designed by human beings with their own inherent preferences and dislikes, often left unarticulated but manifested through the technologies they design, build and increasingly, train. How refreshing a message for a government who has been required to bend their rules to meet the expectations of those whom they want to do business with. Delivered by a man and corporation that has achieved global recognition and success. Completely unlike the MSC where discussions with the international advisory panel then indicated that foreign (read Western) corporations wanted local practices of employment, business ownership and yes, even media governance to be adjusted to their expectations, their view of the world as it should be, before they would invest in and participate in the MSC. What Alibaba, Ma and China now offer are unconditional partnerships with proven models of success that defy the usual 'aid-with-strings' offer of the West. What more could a receiving nation want?

No surprise then that even after a change of government, when the Pakatan Harapan opposition come into power with nonagenarian PM

Mahathir at the helm again, China's wooing of Malaysia did not falter. In fact, Alibaba's Malaysia office in Kuala Lumpur was declared open by Ma in 2018. In his speech at the launch, Jack Ma emphasized: "The difference between our model and American model or solution to globalised model is that we come here looking for partners, we come here to enable partners. We want partners to be local kings here" (Ma 2018). Perhaps Ma had heard that PM Mahathir, who had returned to power again, has a history of anti-Westernism that he could tap into. Or perhaps he is echoing China's official BRI discourse: "adhering to the Path of Peaceful Development and Constructing a Community of Common Destiny with Mankind" (Mardell 2017). The core of the idea is simple—Come, our way of doing business doesn't worry about human rights, privacy or security, it's just plain old business—a seductive proposition for any country that's been stung in the past by capitalist libertarianism and a leader/ship still anti-Western at heart. Not that the Malaysian government was unaware of the debt-traps other developing nations such as Sri Lanka have fallen into with BRI projects financed under terms that the recipient nation could not realistically service (Fägersten and Rühlig 2020). Two months into government the Ministry of Finance suspended three of these infrastructure project associated with the BRI. The projects included the East Coast Rail Link, the Multi Project Pipeline running from Malacca to Kedah, and the Trans-Sabah Gas Pipeline stretching between Kimanis, Sandakan and Tawau, which together came up to a total cost of USD22 billion (Malhi 2018, p. 718). Already during the lead-up to the 14th General Election Mahathir had referred to the BRI projects as a threat to national sovereignty. When the projects were suspended after the PH government came into office, much was made of Mahathir's claims of Chinese neo-colonialism both before and after coming into office (Wang 2018; Anderlini 2018). Eventually, after much behind the scenes negotiation, the ECRL was "rebooted" to use the words of Bai Tian, China's ambassador to Malaysia (Bai 2019), but in payment terms and scale of construction much more modest and manageable. At the same time, plans for a new USD1 billion artificial intelligence hub, AI Park, were approved. A project between Sense-Time, China's AI unicorn, local Malaysian firm, G3 Global and China Harbour Engineering Company (CHEC), the park would comprise a combination of infrastructure construction and technology transfer (G3 Global and SenseTime 2019). The plans for the AI Park do not differ much from how the MSC was envisioned, only the work of and cost for both construction and technology transfer would be the responsibility of foreign investors rather than the government. In one instance PM Mahathir has put his anti-Westernism on display, by snubbing the United States and endorsing Huawei's 5G technologies publicly stating that Malaysia "will use Huawei's gear "as much as possible" because they provide "tremendous advance over American technology" (McCombs 2019). So far a whole host of China-based digital technologies, corporations and technology transfer initiatives have now been welcomed into Malaysia. These range from artificial intelligence in the service of traffic management to relieve congestion i.e. Alibaba's City Brain project (Xiao 2018), smart city technologies and providers, Huawei's Global Training Centre (Sukumaran 2019). and the Malaysia Tianchi Big Data Program (Bragg 2018). Modern history has documented many instances where technologies such as construction engineering and camera film have been designed to exclude certain groups or advantage one group over others (Winner 1989; Roth 2019). Researchers today have revealed that similar issues dog digital technologies. Facial recognition, for example, has facilitated racial profiling and, data-centric management through the use of algorithms has created even greater inequalities between employers and workers (Garvie and Frankle 2016; Mateescu and Nguyen 2019). Despite this, few doubts have been raised about the economic advantage these digital technologies will bring to the nation, let alone critiques of how exclusionary they may be or if IG may be needed to maintain some transparency, fairness or accountability. With the (impending) entry of all these digital technologies, IG in Malaysia can be said to be poised at a turning point.

INTERNET SOVEREIGNTY

Why is Malaysia's turn to Chinese digital technologies and corporations important in terms of this book? In the introductory chapter we alluded to the two-horse race that most pundits believe IG to be. That is to say, the two models that nations have to choose between are the US-centric multistakeholder, distributed model that has sufficed so far and the multilateral, centrist model championed by China. The global digital corporations based in the U.S. have a decided preference for the multistakeholder model, not least because it then makes them no more than just another stakeholder. The multilateral model exemplified by China's notion of Internet sovereignty is, a traditional, sovereignty-based communications governance regime in the international arena. ... an international regime organized around treaty-based intergovernmental organizations that rely on one-country, one-vote distributions of power" (Mueller 2012, p. 181). At its core, Internet sovereignty is an attempt "to align the potent international legal norm of state sovereignty with a key authoritarian priority: absolute control by the regime over the digital experience of its population" (McKune and Ahmed 2018, p. 3835). China's "rhetorical and monetary support" (p. 3850), bolstered by its internet champions' successes as platforms, technological innovators and star performers on the stock markets of the world makes the multilateral model attractive. The number of successful digital corporations based in China are growing and changing constantly. Most recently they include the runaway popularity of TikTok (short video platform owned by ByteDance), the prevalent use of Yitu's facial recognition algorithms and the appeal of Oppo mobile phones on price point alone, to name a few. All of these firms have survived the rough and tumble of doing business in China but more importantly, they have also managed to flourish despite the iron grip that the Beijing has over how the internet is governed throughout the country. The Cybersecurity Law of China came into effect on 1 June 2017. The remit of this law goes hand-in-hand with the broad and ever expanding scope of cyber borders in China, which according to McKune and Ahmed, include: "Internet infrastructure, wireless technology facilities, and cloud computing infrastructure ... Chinese Internet's domain names and related public services" ... as well as "the financial, telecommunications, transportation, energy, and other ... core national networked systems" (McKune and Ahmed 2018, p. 3838).

For semi-authoritarian regimes like Singapore and Malaysia, the notion of internet sovereignty and multilateralism is highly seductive. After all, if nations rather than stakeholders of all shape and form, are the basic unit round which IG is built, then governments are the arbiters of what is legal or not within the country's cyber borders. Further, China's digital unicorns and their successes despite the illiberal conditions of operation within China argue that liberalism and capitalist prosperity are not irreparably intertwined. Ergo, economic prosperity can be achieved within limited political liberty. Hence the welcome of China's digital corporations into Malaysia poses a dilemma for the West because it lends further legitimacy to China's narrative and negates the strength of the narrative that underlines the West—that freedom is intrinsic to personal success.

Since Mahathir is once again at the helm of Malaysia, it is worth recalling how Prime Minister envisioned information technology (IT) would inflect Malaysians' way of life (Mahathir 2002, pp. 156–157):

> We have a clear vision for Malaysia called Vision 2020, the purpose of which is to attain developed-nation status by the year 2020. We want to become a developed nation in our own mould. Malaysia's IT agenda defines the content of the mould as the creation of a civil society. By "civil society" we mean a community which is self-regulating and empowered through the use of knowledge, skills and values inculcated within the people. Such a society will allow every Malaysian to live a life of managed destiny and dignity, not just in the here and now, but also in the future.

The above excerpt is taken from a speech, "Inventing Our Common Future", Mahathir delivered in 1997 in the lead-up to and surrounding the MSC. While Xi Jinping's invocation of a future of common destiny seems to echo this call, Mahathir's depiction of a self-regulating and empowered civil society in Malaysia is the polar opposite of what China intends with internet sovereignty and public opinion guidance. This is especially so when we consider how BoG7 and the political changes it facilitated has left its mark on how Malaysians understand the connection information flow, enfranchisement and democracy. In fact, the flow of information via the Internet is a big part of how the current *Pakatan Harapan* government came into power. Whereas, as McKune and Ahmed point out, "the CCP's state centric approach to the Internet stems from long-standing concerns over regime control and domestic stability, both of which information flows deeply affect" (McKune and Ahmed 2018, p. 3838). It is true the potency of BoG7 has been diluted with the caveat mentioned earlier. In Malaysia, at least, it is not quite the Wild West of the early Internet described by Barlow (1996) but there remains sufficient latitude and flexibility in its hybrid IG model to make it a bad fit with internet sovereignty à la China.

Private Platforms, Public Oversight

The interweaving of socio-political events and changes in how the internet is governed here is deliberate. The point being that IG at the local level cannot be understood without taking into account the local socio-political context. Even so, understanding the context of an act of governance i.e.

the politicisation of IG, does not negate its intent. While enacting laws to stifle news of a public official's misdeeds and shore up their political position is morally indefensible, the spread of rumours and misinformation can cause irreparable damage. More corrosive still is information uploaded by malicious parties or outdated information of one's youthful discretions retained online, accessible to all and sundry. The European Union's 'Right to be Forgotten' (also known as Right to Erasure, hereafter RTE) tucked under the General Data Protection Regulation (GDPR) as Article 17, seeks to return control of user data to individual users' hands (The Information Commissioner's Office, n.d.). Of course, there are limitations to both how soon and, how far the erasure can be taken as well as the costs to businesses and institutions involved in such erasure but there are other issues with this and other similar regulations. Firstly, it leaves the decision as to what should or should not remain online rather than erased to corporations like Google and Facebook. As Suzor reminds us (2019, p. 37):

> the rules that social media platforms develop are not democratically selected nor subject to review by an independent judiciary that is empowered to strike down rules that are unfair or overly restrictive. The content moderation process is human and fallible, like real laws are, but it is not surrounded by an infrastructure designed to lessen the influence of individual decision-makers and the chance of bias.

In other words, the discretionary power exercised by these platforms as intermediaries are "lawless". Secondly, what laws such as RTE do is hand the responsibility for how, who, when, why and where data is retained back to individuals who have to put in the request to corporations like Google. The tacit assumption is that individuals comprehend and can deal with the bureaucratic complexities involved in placing any such request in the first place. This effectively limits the RTE to those who are able and willing to tackle the system. Currently, the most popular platforms in the world, owned by some of the largest digital corporations in the world, hail from the West, which leaves far too much of the power as to how, where, when, why and what behaviours are permissible online to these platforms. Even greater transparency on the part of platforms, often advocated as a solution to the impasse is a far from ideal solution (Ananny and Crawford 2016). Current arguments about IG have been locked at the impasse of platform governance. It is true platforms should not have the power to dictate how human lives unfold but it is the case because these platforms are now so inextricably entwined with daily lives, social and labour functions they are well nigh indispensable for large swathes of the world's populations. Take the common task of finding the way from point A to point B. In the past, street directories, maps and directions from others would be how an individual would make their way to point B. Such methods for locating onself and others is now more the exception than the rule. Replaced first with guidance from the skies via geographic positioning Systems (GPS) and then, with the mass take-up of smartphones, even GPS itself has been replaced by Google Maps, Waze and any number of way-finding apps. The prevalence of these apps was gradual but ask directions of anyone these days and most would whip out their smartphones. The same can be said of how phone numbers, addresses and contacts are stored, called up and used. Where recalling familiar and important phone numbers and addresses used to be a norm, mobile internet coupled with smartphones have replaced the exercise of human memory. Inevitable or not, digital platforms now mediate how most humans communicate, travel, do business, work and socialise. It is, as mentioned in Chapter 2, precisely the Internet's ubiquity that makes IG a pressing issue for all.

Suzor argues that since these platforms are making decisions that could affect the fundamental rights of individuals "we are entitled to expect a greater degree of legitimacy" (Suzor 2019, p. 165). His suggestion is the formation of a global digital constitution but he stops short of detailing how such a constitution would be arrived at or who would be invited to the decision-making table, so to speak. Or perhaps more importantly, who would not be. Even assuming that all interested stakeholders do gather, reaching consensus on these issues are never straightforward. In 2018 this author had the opportunity to witness how tricky such consensus building exercises can be. In August 2018 approximately 3 months after the Pakatan Harapan government came into office, the Centre of Independent Journalism (CIJ) and Malaysian Centre for Constitutionalism and Human Rights (MCCHR) sent out an invitation to all stakeholders to gather at a local hotel to hammer out what the terms, responsibilities, remit and makeup of a Malaysian Media Council might be. Over two days the hot-housed mix of journalists, editors, bureaucrats, academics and members of non-government organisations such as the UN, the CIJ and the Ministry of Information from Peninsular and Borneo Malaysia engaged in vigorous argument, workshopping ideas, gathering views and trying to pin down the basic form of a Malaysian Media Council. Despite the goodwill expressed in their attendance, it gradually became clear that without addressing the grievances of those disadvantaged or wronged in the past and merely papering over, for example, the distinctions that have emerged between media governance, conditions and expectations in Borneo and Peninsular Malaysia, moving ahead would remain an ideal. The same can be said of the plight of a populace whose education has been a political football over the decades. Despite the place of English as the lingua franca in Southeast Asia, the medium of instruction for science and maths in Malaysia's public education system has been a matter of debate, switching between Bahasa Malay and English as various interests lobby and win over the government (Chan and Tan 2006; Dr M wants 2020). While not historical grievances as such, the issue speaks firstly, to the need to ready a population to receive the knowledge and technology transfers that digital technologies introduced into Malaysia are intended to produce; but more importantly, for the nation to continue finding its own stance on the socio-political, cultural and economic changes that digital technologies in the form of software, hardware, infrastructure, user patterns and constraints will bring to ordinary Malaysians' daily lives. Without the right training, Malaysians stand to become unwitting pawns again.

This brings us back to the task of global internet governance. As laid out in Chapter 2 and here, where the newly formed nation-states of Singapore and Malaysia were inclined to adopt the legal legacies of their former colonial administrator, both have, in the matter of IG proven more than equal to the task of forging their own paths. Mahahtir's intransigent anti-Westernism aside, the terms 'global and globalisation' have never really been as all encompassing as they suggest, rather they have traditionally alluded mainly to the West. The East or to use a more contemporary term, the Global South has seldom been party to global decision-making processes. That is why in its advocacy for internet sovereignty China has consistently referred to the bogeyman of the internet hegemon and acted as if it was doing so on behalf of the Global South on matters IG (McKune and Ahmed 2018). More broadly, China's BRI and its associated project, the Digital Silk Road, routinely resorts to the counter-hegemonic discourse of an "open and inclusive globalisation" (Seoane 2019). as it makes its way across various developing nations. None of this is lost on the nations of the Global South who have collectively (Kaul 2013, p. 2):

become increasingly active participants in global policymaking. Moreover, they have used their influence not only to further their own, narrowly defined national interests but also to shape the normative framework of global policymaking by emphasizing concerns like enhancing the fairness and justice of international negotiations or foster a better balance between growth and development as well as public and private interest.

One way, then, to read Malaysia's openness to China's digital technologies and corporations such as Huawei and Alibaba is as portrayed by media i.e. a snub to Trump's United States and an extension of the Prime Minister's entrenched anti-Westernism. The other possibility is to understand the growing role of the Global South in global internet governance. However these gestures are read, what is clear is that China's increasing influence in contemporary Southeast Asia is a factor that will more heavily influence the hybrid model of internet governance that Malaysia adapts and adopts in the coming decades. While Southeast Asian countries like Singapore and Malaysia are often painted as being part of the backdrop against which the theatre of East-West/US-China tensions are played out, it is as well to recall that today neither Singapore nor Malaysia are passive recipients of or bystanders to the machinations, diplomatic proposals and claims of suzerainty mentioned above but full-fledged nation-states with their own views, agendas and interests to guard and advance.

BIBLIOGRAPHY

Ananny, M., & Crawford, K. (2016). Seeing Without Knowing: Limitations of the Transparency Ideal and Its Application to Algorithmic Accountability. *New Media and Society, 20*(3), 973–989.

Anderlini, J. (2018). *China Is at Risk of Becoming a Colonial Power.* https://www.ft.com/content/186743b8-bb25-11e8-94b2-17176fbf93f5. Accessed 10 July 2020.

Attorney-Governor's Chambers. *Anti-Fake News Act 2018.* www.federalgazette.agc.gov.my/outputaktap/20180411_803_BI_WJW010830%20BI.pdf. Accessed 10 July 2020.

Bai, T. (2019). *BRI Wind May Just be What Malaysia Needs.* https://www.nst.com.my/opinion/columnists/2019/04/482008/bri-wind-may-just-be-what-malaysia-needs. Accessed 10 July 2020.

Barlow, J. P. (1996). *A Declaration of the Independence of Cyberspace.* https://www.eff.org/fr/cyberspace-independence. Accessed 2 March 2020.

Bragg, T. (2018). *How Malaysia Plans to Win the Smart City Race*. https:// techwireasia.com/2018/01/malaysia-plans-win-smart-city-race/. Accessed 10 July 2020.

Chan, S. H., & Tan, H. (2006). English for Mathematics and Science: Current Malaysian Language-in-Education Policies and Practices. *Language and Education, 20*(4), 306–321.

Dr M Wants Science and Maths to be Taught in English again. (2020). https:// www.malaysiakini.com/news/509049. Accessed 10 July 2020.

Fägersten, B., & Rühlig, T. (2020). *China's Standard Power and Its Geopolitical Implications for Europe*. https://www.ui.se/globalassets/ui.se-eng/publicati ons/ui-publications/2019/ui-brief-no.-2-2019.pdf. Accessed 10 July 2020.

G3 Global and SenseTime to set up the First Artificial Intelligence Park in Malaysia. (2019). http://www.g3global.com.my/news_aipark.html. Accessed 10 July 2020.

Garvie, C., & Frankle, J. (2016). *A Facial-Recognition Software Might Have a Racial Bias Problem*. https://www.theatlantic.com/technology/arc hive/2016/04/the-underlying-bias-of-facial-recognition-systems/476991/. Accessed 10 July 2020.

Hope, B., & Wright, T. (2018). *Billion Dollar Whale*. New York: Hachette.

Jack Ma and PM at DFTZ dialogue. (2017). https://www.youtube.com/watch? v=YlcfYVqagKU. Accessed 3 November 2019.

Jash, S. (2019). *The Creeping Rise of India's Internet Shutdowns*. https:// www.newamerica.org/weekly/edition-244/creeping-rise-internet-shutdowns-india/. Accessed 11 April 2019.

Hopkins, J. (2014). Cybertroopers and Tea Parties: Government Use of the Internet in Malaysia. *Asian Journal of Communication, 24*(1), 5–24.

Kaul, I. (2013). *The Rise of the Global South: Implications for the Provisioning of Global Public Goods* (UNDP-HDRO Occasional Papers No. 2013/08). https://ssrn.com/abstract=2344483. Accessed 10 July 2020.

Koya, Z. (2018). *Bilqis Hijjas: When I Dropped the Yellow Balloons on Najib...*. https://www.awanireview.com/articles/2018/08/23/news/bilqis-hijjas-when-i-dropped-the-yellow-balloons-on-najib-440/. Accessed 10 July 2020.

'Lawyer Haris Ibrahim gets eight months' jail for sedition. (2016). https://www. malaysiakini.com/news/337770. Accessed 10 July 2020.

Leong, S. (2018). Prophets of Mass Innovation: The Gospel According to BAT. *Media Industries, 5*(1), 69–87.

Lunden, I. (2019). *Iran Shuts Down Country's Internet in the Wake of Fuel Protests*. https://techcrunch.com/2019/11/17/iran-shuts-down-cou ntrys-internet-in-the-wake-of-fuel-protests/. Accessed 17 November 2019.

Ma, J. (2018). *Jack Ma's Full Speech During KL Alibaba Office Launch*. https:// www.youtube.com/watch?v=nUrQq3VU3Gc. Accessed 18 June 2019.

Mahathir, M. (2002). Inventing Our Common Future. In M. Bin Mohamad, D. N. Abdulai, T. C. Ng, & N. T. Chuan (Eds.), *Mahathir Mohamad: A Visionary & His Vision of Malaysia's K-Economy* (pp. 153–162). Subang Jaya: Pelanduk.

Malhi, A. (2018). *Race, Debt and Sovereignty—The 'China Factor' in Malaysia's GE14.* https://www.commonwealthroundtable.co.uk/commonwealth/eur asia/malaysia/race-debt-and-sovereignty-the-china-factor-in-malaysias-ge14/. Accessed 10 July 2020.

Malaysia Anti-Fake News Act. (2018). https://www.article19.org/resources/mal aysia-anti-fake-news-act/. Accessed 24 April 2018.

Malaysia's National eCommerce Strategic Roadmap. (2016). https://issuu. com/peerasakchanchaiwittaya/docs/malaysias_national_ecommerce_str ate/15. Accessed 10 July 2020.

Mardell, J. (2017). *The 'Community Of Common Destiny' in Xi Jinping's New Era.* https://thediplomat.com/2017/10/the-community-of-common-destiny-in-xi-jinpings-new-era/. Accessed 10 July 2019.

Mateescu, A., & Nguyen, A. (2019). Explainer: Algorithmic Management in the Workplace. https://datasociety.net/pubs/algorithmic-management-explainer. pdf. Accessed 10 July 2020.

McCombs, D. (2019). *Malaysia's Mahathir Backs Huawei in Rare Public Rebuke of U.S.* https://www.bloomberg.com/news/articles/2019-05-30/ mahathir-backs-huawei-in-rare-public-rebuke-of-u-s-clampdown. Accessed 10 July 2020.

McKune, S., & Ahmed, S. (2018). The Contestation and Shaping of Cyber Norms Through China's Internet Sovereignty Agenda. *International Journal of Communication, 12,* 3835–3855.

MCMC Annual Report 2009. https://www.mcmc.gov.my/skmmgovmy/files/ ff/ffb57736-1dff-461b-9a52-d85f25631e86/index.html#19/z. Accessed 10 July 2020.

MCMC Annual Report 2013. https://www.mcmc.gov.my/skmmgovmy/ media/General/pdf/MCMC-AR_ENG_2013.pdf. Accessed 10 July 2020.

MCMC Annual Report 2018. https://www.mcmc.gov.my/skmmgovmy/ media/General/pdf/HPUS2018.pdf. Accessed 10 July 2020.

MCMC Handphone Users Survey 2014. https://www.mcmc.gov.my/skm mgovmy/media/General/pdf/MCMC-Hand-Phone-User19112015.pdf. Accessed 10 July 2020.

MCMC Summons 15 over spreading of rumours. (2013). https://m.malaysiak ini.com/news/228657. Accessed 10 July 2020.

MIDA (Malaysian Investment Development Authority). (2019). https://www. mida.gov.my/env3/uploads/IncentivesCompilation/MDEC/2013/AppII. pdf. Accessed 10 July 2020.

Mueller, M. L. (2012). China and Global IG: A Tiger by the Tail. In R. Deibert, J. Palfrey, R. Rohozinski, & J. Zittrain (Eds.), *Access Contested: Security, Identity, and Resistance in Asian Cyberspace* (pp. 177–194). Cambridge, MA and London: MIT Press.

Noor, F. A. (2013). The Malaysian General Elections of 2013: The Last Attempt at Secular-inclusive Nation-Building? *Journal of Current Southeast Asian Affairs, 32*, 2.

Razak, A. (2011). *Government Ready to Contravene MSC's Bill of Guarantees.* https://www.malaysiakini.com/news/154562. Accessed 10 July 2020.

Rewcastle-Brown, C., & Brown, G. (2018). *The Sarawak Report.* London: Lost World Press.

Roth, L. (2019). Looking at Shirley, the Ultimate Norm: Colour Balance, Image Technologies, and Cognitive Equity. *Canadian Journal of Communication, 34*(1), 111–136.

Seoane, M. F. V. (2019). Alibaba's Discourse for the Digital Silk Road: The Electronic World Trade Platform and 'Inclusive Globalization'. *Chinese Journal of Communication, 13*(1), 68–83. https://doi.org/10.1080/17544750.2019.1606838.

Sukumaran, T. (2019). *Malaysia Welcomes Chinese Tech Giant Huawei Despite Western Concerns.* https://www.scmp.com/week-asia/geopolitics/article/3006262/malaysia-welcomes-chinese-tech-giant-huawei-despite-western. Accessed 10 July 2020.

Suzor, N. P. (2019). *Lawless: The Secret Rules That Govern Our Digital Lives.* Cambridge: Cambridge University Press. https://doi.org/10.1017/9781108666428.

The Information Commissioner's Office. (n.d.). *The Right to Erasure.* https://ico.org.uk/for-organisations/guide-to-data-protection/guide-to-the-general-data-protection-regulation-gdpr/individual-rights/right-to-erasure/. Accessed 10 July 2020.

Vee, V. T. (2011). The Struggle for Digital Freedom of Speech: The Malaysian Sociopolitical Blogsphere's Experience. In R. Deibert, J. Palfrey, R. Rohozinski, & J. Zittrain (Eds.), *Access Contested: Security, Identity, and Resistance in Asian Cyberspace.* http://access.opennet.net/wp-content/uploads/2011/12/accesscontested-chapter-03.pdf. Accessed 1 February 2020.

Wang, X. (2018). *Mahathir'S Pushback Shows China's Belt and Road Plan Needs Review.* https://www.scmp.com/week-asia/opinion/article/2161069/mahathirs-pushback-against-chinese-deals-shows-belt-and-road-plan.

Whiting, A. (2011, December 6). Malaysia—Assembling the Peaceful Assembly Act. *New Mandala.* https://www.newmandala.org/malaysia-assembling-the-peaceful-assembly-act/. Accessed 10 July 2020.

Winner, L. (1989). *The Whale and the Reactor: A Search for Limits in an Age of High Technology* (pp. 19–39). Chicago: University of Chicago Press.

Xiao, E. (2018). *Alibaba Doubles Down on Malaysia, Rolls Out Traffic Control System.* https://www.techinasia.com/malaysia-city-brain. Accessed 10 July 2020.

Yeung, J. (2018). *Malaysia Repeals Controversial Fake News Law.* https://edition.cnn.com/2018/08/17/asia/malaysia-fake-news-law-repeal-intl/index.html. Accessed 10 July 2020.

CHAPTER 5

Internet Governance: Singapore's Regulatory Influence

Abstract This chapter details the governance mindset that the Singapore government has applied to managing and mitigating the risks of open internet access. Apart from cyber-security, the authoritarian People's Action Party (PAP) government in Singapore was—and still is—most concerned about ensuring that it retains political control over its citizens. The approach taken by Singapore authorities has been premised on ensuring that commerce would thrive by embracing the internet and digital technologies, but its collateral, the free-for-all libertarian ideas that accompany an open cyberspace, would be curtailed via the Foucault-inspired discourse of 'auto-regulation' that enables surveillance via balanced, light-touch regulatory principles. The chapter concludes by positioning internet governance in Singapore vis-à-vis the nation's delicate foreign policy approaches. It argues in the final analysis that the Internet is intrinsically hybrid, and that only by thinking about local applications and controls whilst simultaneously recognising the global technology that the Internet represents can one govern the Internet sufficiently, and with potency.

Keywords Internet regulation · Foucault · Governmentality · Balanced and light-touch · Auto-regulation · Self-regulation · Panopticon

INTRODUCTION: SINGAPORE'S GOVERNANCE SETTING

[T]he internet is being and will be regulated. It is too important a medium for communication, entertainment and education not to be regulated. The libertarian approach of no-regulation on the net is just unsustainable. To believe that the internet cannot be regulated would lead one not to promulgate laws for the internet. Not to have laws will, in effect, undermine the development of the developing world and risk them being left further behind. However, it does not mean that regulations are bad in themselves. It is the quality, not the quantity, of the regulations that matter. (Ang 2005, p. 14)

In December 1996, the United Nations Educational, Scientific and Cultural Organisation (UNESCO) commissioned the Australian Broadcasting Authority (ABA), the predecessor to the present Australian Communications and Media Authority (ACMA), to conduct a pilot study to consider a range of issues pertaining to the increasing spread of the internet as a communications and information medium. The final report published in October 1997, entitled *The Internet and some international regulatory issues relating to content* (ABA 1997), provided a comparative overview of the regulatory developments in four countries: Australia, the United Kingdom, Malaysia and Singapore. The main reasons why these four countries were chosen was that they were seen not only as early adopters of mass public Internet use, but were also some of the earliest to either pass legislations, or flag their intentions to, related to Internet use and regulations outside of the United States.

This early study on Internet governance revealed the advanced state of Internet use as well as the rapid development of a policy framework to regulate the Internet in the city-state of Singapore relative to the other three countries surveyed. In particular, the ABA report noted that Singapore had declared from the early days of public Internet access that it planned to identify ways to control Internet traffic, but at the same time, was adamant that it did not intend to over-regulate the Internet (ABA 1997, p. 39). While this report was well known by governments and regulators around the world, it did not make headline news in most places because the dominant orthodoxy at the time was that the Internet was a free medium that transcended political censorship and control. As an open source technology designed and founded on libertarian ideology and cultural conceptions of freedom, the advent of the

Internet brought with it hopes of democratic people power to places that have been socially, economically and politically disenfranchised by dictatorial or authoritarian governments (Castells 2001, pp. 31–33). But like Malaysia, Singapore had other ideas about Internet governance. As Professor Ang Peng Hwa—one of the earliest Internet and communication scholar to have expounded on the rationales *for* global Internet governance and regulation—has noted (in the opening quote above), Singapore took the position that the Internet was too important to let loose (Ang 1999, 2005).

Recognising that it is the quality, as opposed to the quantity, of Internet regulations that mattered (Ang 2005, p. 14), the Singapore government opted for a calibrated approach towards Internet governance. It was clear from the outset that the widely held economic potential of the Internet had led Singaporean authorities to opt for a light—or less restrictive—approach towards managing the Internet, as opposed to the strict and censorious 'lockdown' of traditional print and broadcast media that Singapore had become notorious for by the 1990s (Lee 2010). What this 'light' and 'calibrated' approach entailed was by no means clear, not least because the technology, and its capabilities, was still in its infancy. What was clear though was that Singapore had strong regulatory tendencies because it viewed social, cultural and policy control of information and media spaces as critically important to its political sustenance and sustainability (Lee 2010).

In hindsight, it is now clear that Singapore was in fact publicly articulating, albeit without the same language that is used in more contemporaneous times, its faith in the principles of Internet sovereignty. This belief is played out in regulatory and policy terms that are, and have been, consistent with how the Singapore government has conducted media governance, which is that media, information and all modes of communication must be beholden to the authoritarian values of the state (Lee 2016, p. 82). As anthropologist and Singapore-watcher Yao Souchou clarified back in 1996, using the term 'techno-triumphalism' to describe Singapore's desire to extend geophysical sovereignty into the borderless world of the internet:

> The State of Singapore has always seen itself as having the legitimate right to influence and manage citizens' choices in a wide range of activities from what one can read or watch on the screen and television to the right to chew gum, and the Internet is no exception. (Yao 1996, p. 73)

Indeed, Yao's 'prophetic' statement about the prospect of Singapore leading the way in Internet governance and sovereignty, made at the dawn of public Internet roll-out in the mid-1990s, was to be validated shortly afterwards. Singapore's so-called 'light touch' regulatory regime for the Internet was unveiled as a matter of course in July 1996—and continues to be a major component of Singapore's Internet governance regime for the most part as we enter the fourth (and fifth) decade of public Internet use.

The next section of this chapter unpacks this 'light touch' regulatory approach, which forms the precursor to understanding latter governance framework introduced by the Singapore government. The light-touch regulation leads us to an Internet class licensing scheme which automatically makes all users of the internet liable to a set of rules. The class licence scheme is complemented by a Schedule entitled 'Internet Code of Practice', intended as industry guidelines for Internet Service Providers and Internet Content Providers. Both of these codes, developed in consultation with key regulatory partners and industry players, are intended to encourage responsible use of the new medium whilst facilitating its growth and development in Singapore. Both codes combine to form Singapore's self-regulatory Internet policy relating to content, which the government describes as 'balanced and light-touch' in view of its tacit acceptance by the Internet industry and what might be perceived as a light-handed, or even a hands-off, approach taken by the regulator.

This chapter also highlights how the government has also strategically mobilised 'gestural' practices such Internet filtering and a constant 'blacklisting' of selected sites to demonstrate that it is (pro)actively governing the Internet. One of the most cited examples is in the government's decision to ban a token 100 pornographic sites to showcase to the public the nation's conservative values (Lee 2010). In addition, on the policing front, cyber-surveillance of 'criminal sites' and the 'dark web' by the Singapore Police Force is carried out to not just to ward of cybercrime, but also to send 'chilling messages' to users that the Internet is not fully safe, can be monitored and will indeed be monitored (ibid.).

As Singapore enters the Smart Nation era, it has become a prime example of a 'panoptic' system of inert yet visible internet surveillance through its approach towards internet governance and control. This is what co-author Terence Lee has described as the Singaporean discourse of Internet or technological auto-regulation (see Lee 2002, 2005, 2010;

Lee and Birch 2000), which is based on Foucault's works on governmentality and surveillance. Auto-regulation governs the everyday conduct of individual and corporate users of the Internet by stealth, but utilises other social, cultural, economic, political and indeed legal apparatuses to ensure compliance and control.

In 2019, Singapore passed a new law—the Protection from Online Falsehoods and Manipulation Act (POFMA)—that grants additional power to the government to issue correction orders to what it deems as fake news, known euphemistically in Singapore as "deliberate online falsehoods" (Lee and Lee 2019). This new law requires an online user to carry a correction notice alongside the 'fake news' post or article immediately if a government minister determines that it is erroneous. The invocation of POFMA in the name of 'public interest' continues a history of consistency in the application of Internet auto-regulation and governmentality in Singapore. To the extent that Internet governance in Singapore has become a reference point to how one can manage, or even rein in, oppositional, seditious or simply socially and politically problematic voices and activities online without sacrificing financial/commercial success (Luger 2020). It is therefore not difficult to see how and why Singapore's internet auto-regulation governance approach has become a workable model of a successful authoritarian model of internet management and control, especially since it is also one that emphasises—at its core— important principles of supranational juridical rights alongside national sovereignty. In other words, Singapore sets itself up to embrace both multistakeholderism and multilateralism simultaneously, choosing in the process not to view them in contradictory terms. Differently put, Singapore's Internet governance framework in many ways gives us a glimpse of what a hybrid model might possibly look like.

The Light-Touch Principle

The term 'light-touch' (with or without the hyphen) typically refers to "a situation in which only a few people are in charge of something or something is not controlled very strictly".[1] It even gives the impression of something being attended to in a relaxed, friendly, even humourous, manner. When applied to policy or in a regulatory context, the concept

---

[1] https://dictionary.cambridge.org/dictionary/english/light-touch (accessed 14 April 2020).

of 'light-touch' usually points to self-regulation on the part of the user and/or producer based on an agreed-upon code or guidance. While Singapore has used the term 'light-touch' in reference to its approach to Internet regulation, it has more recently opted for the terms 'pragmatic' and 'balanced', as captured in the Internet Regulatory Framework section of the Info-Communications Media Development Authority's (IMDA) website:

> Singapore is one of the most-connected countries in the world, with high broadband and mobile phone penetration. IMDA is mindful of the dynamic and borderless nature of the Internet and the need for the responsible use of this medium. In light of this, IMDA adopts a pragmatic and balanced approach in regulating the Internet.[2]

On paper (and in government information websites), Singapore adopts and embraces a similar light-touch or 'balanced' principle in relation to the Internet. This does not mean though that, in practice, the Singaporean authorities are forceful or high-handed. Rather, the application—and indeed, the successful roll-out—of Singapore's light-touch principle towards Internet governance relies heavily on what Terence Lee has referred to as "auto-regulation" (Lee, 2002, 2005, 2010). Auto-regulation, which borrows strongly on aspects of French philosopher Michel Foucault's neoliberal discourse of 'governmentality', occurs when Internet users are steered towards making 'correct' choices and decisions via the joint application of overt legislative codes with other less-overt mechanisms of 'discipline' (Foucault 1977). Foucault's notion of 'discipline' is understood in this context as a functional device or apparatus aimed at making the exercise of power more efficient through the pragmatic but no less subtle coercion of people as both cultural citizens and economic subjects (Foucault 1977, pp. 136–8).

A 'light-touch' self-regulatory principle has worked—and continues to work—in Singapore not because of its functionality or the inherent ability of individuals or Internet industry players to discipline or conduct themselves. After all, laws are often arbitrary to non-legal experts, which is what the clear majority of Internet users are. Rather, the practicability of

---

[2] https://www.imda.gov.sg/regulations-and-licensing-listing/content-standards-and-classification/standards-and-classification/internet (accessed 15 April 2020).

self-regulation relies on the application of auto-regulation, where policies and legal codes are deemed as governmental tools and technologies employed by regulatory authorities to 'shape, normalize and instrumentalize the conduct, thought, decisions and aspirations of others' (Miller and Rose 1990, p. 82). With the holistic application of 'auto-regulation', the otherwise complex and arduous tasks of Internet governance and policing in Singapore are made less onerous as they are aided, empowered and 'co-regulated' by laws, policy codes, statements and technologies of surveillance designed to shape the conduct of economic individuals and groups within society.

In 1994, the Singapore Broadcasting Authority (SBA) Act—subsequently referred to as the Broadcasting Act (1994)—was established in Singapore, giving it the full mandate to manage the nation's Internet policy and regulate Internet content. In 2003, the newly-established Media Development Authority of Singapore (MDA) replaced the SBA, and thus took over the internet governance role. With increasing convergence of the media, IT and telecommunications sectors, MDA was further consolidated and reformed into the Info-Communications Media Development Authority (IMDA) in October 2016. Not unlike large regulatory agencies in other jurisdictions, such as the Federal Communications Commission (FCC) in the United States of America or the Australian Communications and Media Authority (ACMA) in Australia— the IMDA's role is multi-faceted, of which Internet governance is but one of its critical roles.[3]

The IMDA website makes it known that the Singapore government takes a 'light-touch' approach towards regulating the Internet:

> Our key focus is on content issues of concern to Singapore such as those relating to public interest, race, religion, pornography and content harmful to children. As a symbolic statement of our societal values, local ISPs are required to restrict public access to a limited number of mass impact websites which contain content that the community regards as offensive or harmful to Singapore's racial and religious harmony, or against national interest. The majority of the websites on the list are pornographic in nature. IMDA does not restrict or monitor individuals' access to online

---

[3] Information about the FCC and ACMA can be accessed from their websites: www.fcc.gov and www.acma.gov.au respectively.

content. Neither does IMDA regulate webpages operated by individuals and personal communications such as email and instant messaging.[4]

Singapore's light-touch principle is predicated upon the 'Broadcasting (Class Licence) Notification' which came into effect on 15 July 1996 (Lee and Birch 2000, pp. 160–161)—and subsequently revised in May 2013, mainly to include online subscription services to the regulation. Commonly referred to as the 'Class Licence scheme', this subsidiary piece of legislation is presented as a convenient and communitarian approach to regulation because it embraces self-regulation on the part of the content provider or host and to a lesser extent the Internet user who participates in online discourses. The 'automatic' Class Licence scheme is complemented by a Schedule entitled 'Internet Code of Practice' (first enacted in July 1996, the current form was revised in November 1997), intended as industry guidelines for Internet Service Providers and Internet Content Providers. Both of these codes, apparently developed in consultation with key regulatory partners and industry players, are intended to encourage responsible use of the new medium whilst facilitating its healthy development in Singapore (ABA 1997, p. 39). Both of these codes—freely available for download on the IMDA website—combine to form Singapore's self-regulatory Internet policy framework in relation to content and the dissemination of information.

Singapore's Internet regulatory framework was described in statements by the former SBA and MDA as 'balanced and light-touch' in view of its tacit acceptance by the Internet industry. As we have identified, while the present preferred language adopted by the IMDA—as well as its parent ministry, the Ministry Communications and Information (MCI)—is 'pragmatic and balanced', the approaches have not changed even as the Internet has morphed from a peripheral technology into the very centre of most economies around the world. Indeed, it is remarkable that both the Class Licence scheme and the Internet Code of Practice have remained largely intact and fully in effect since the mid-1990s, barring a couple of updates to account for new names and new ways of using technologies, such as the prevalence of mobile Internet (IMDA 1997, 2013).

The concept of self-regulation as a means of Internet governance, irrespective of whether it is 'light-touch' or otherwise, appeals to many

---

[4] www.imda.gov.sg/regulations-and-licensing-listing/content-standards-and-classific ation/standards-and-classification/internet (accessed 15 April 2020).

people primarily because it appears to vindicate the liberal ideology which the Internet is said to represent. For the individual and public Internet users, it is a step closer towards having some degree of freedom, whether these come in the social, cultural, economic or political domains. For commercial players operating in a capitalist and consumerist society, self-regulation supports the consumer's or citizen's right to free choice. For government administrators, self-regulation makes the arduous task of monitoring, accounting and reporting less onerous, especially when licensees are 'classed' or categorized collectively, and therefore subject to a uniform set of rules.

The light-touch principle of Internet self-regulation, as applied to the public Internet user, the industry player and the government, is presented in Singapore as a three-pronged approach, namely: instituting a balanced and pragmatics framework, encouraging industry self-regulation and promoting media literacy and cyber wellness through public education. These three approaches are explained in IMDA documentation, framed in simple policy language, as follows:

Instituting a balanced and pragmatic framework: IMDA's regulatory framework for the Internet is embodied in the *Broadcasting (Class Licence) Notification*. Under the Internet Class Licence, Internet Content Providers and Internet Service Providers are deemed automatically licensed and have to observe and comply with the Internet Class Licence Conditions and the Internet Code of Practice, which outlines what the community regards as offensive or harmful to Singapore.

Encouraging industry self-regulation: The industry is encouraged to self-regulate and be socially responsible for their content. IMDA encourages content providers in Singapore to develop industry codes of practice which can be used to promote greater industry self-regulation and complement existing Internet content regulations.

Promoting media literacy and cyber wellness through public education: IMDA recognises the need to educate the public on both the advantages as well as the downsides of the information superhighway. In view of this, IMDA initiates programmes to promote media literacy and the discerning use of the media including the promotion of cyber wellness. To take such efforts further, the Media Literacy Council (MLC) was formed in August 2012 to spearhead public education programmes and initiatives on media literacy and cyber wellness. As its Secretariat, IMDA supports the Council's programmes to build awareness of media and digital literacy issues and promote responsible online participation.[5]

Since the quid pro quo of a class licensing approach to self-regulation is that all agencies involved are compelled to agree to operate responsibly and in accordance with the laws of the land, the fear of legal and/or political reprisals ensures that non-compliance does not occur (Lee and Birch 2000, p. 160). As would be expected, no one in Singapore has yet been taken to task for violating MDA's Internet policy guidelines, although several users have been (fore)warned or admonished for various online activities over the years. The authorities have used this broad adherence to argue favourably for Singapore's excellent Internet governance record. It has also given credit to the third approach of public education and mediation strategies that emphasise media literacy and attention to cyber wellness, which is defined by the Ministry of Education in Singapore as "the positive well-being of Internet users [which] involves an understanding of online behaviour and awareness of how to protect oneself in cyberspace [so as] to become responsible digital learners".[6]

We suggest, however, that such an untarnished record in policy adherence has less to do with notions of regulatory efficacy, governmental transparency or the use of simple policy language such as 'light-touch' or 'balanced'. Rather it has a lot to do with the way Internet governance is conducted via the application of auto-regulation (Lee 2010). Our argument that follows in the next section looks at how the light-touch principle relies on governmental 'auto-regulation' to ensure not just acquiescence or buy-in from the general public, but also support ranging from tacit to explicit.[7] This was particularly pertinent and significant during in the earlier decade (1990s–2000s) of Singapore's Internet history.

---

[5] The three-pronged approach to Internet regulation is also found in the IMDA website as follows: www.imda.gov.sg/regulations-and-licensing-listing/content-standards-and-classification/standards-and-classification/internet.

[6] www.moe.gov.sg/education/programmes/social-and-emotional-learning/cyber-wellness (accessed 16 April 2020).

[7] See Terence Lee (2010, Chapters 5 and 6) for a detailed record of the individuals and groups in Singapore society that have experienced the force of the law in relation to media and Internet misdemeanours in Singapore over the past two decades.

AUTO-REGULATION: DISCIPLINING THE INTERNET

When first unveiled in 1996, two years after mass subscription to the Internet commenced, Singapore's Internet policy did not face much public opposition as concerns at the time were centred on the proliferation of violence and pornographic content. One year later, the former Internet content regulator, the SBA, announced that a shortlist of one hundred pornographic websites were to be blocked via the proxy servers of the three main government-controlled public Internet Service Providers (ISPs): SingTel, M1 and Starhub. This became the first instance of en bloc direct Internet censorship anywhere in the world, and certainly one that prompted both critics and sympathizers of authoritarian rule to cast their eyes on Singapore.

Despite receiving widespread criticisms and scorn, the government's stand on this new policy was firm and unapologetic. According to the authorities, this censorship was an important gesture of concern to bring Internet content in line with the nation's societal and Asian cultural values (Lee 2005). Although this blatant act of censorship offended critics and other liberal-minded citizens, local media support for the move made it publicly acceptable in Singapore in a very short time, notwithstanding the fact that the vast majority of Internet users in Singapore were probably not interested in accessing these banned sites in the first place. The preference for the authorities at the time, nevertheless, was to err on the 'safe' and 'conservative' side, mostly to ensure continued political support from the 'moral majority' of Singaporeans who would readily profess social and cultural conservatism when asked or surveyed. Of course, whether this 'conservatism' is a myth or a representation of reality becomes irrelevant in a climate where cultural control and policing, via the symbolic declaration of the government's and community's values, hinge on the politically expedient enforcement of censorship (Lee and Birch 2000). In any case, the advent of web 2.0 from the mid-2000s led by social media and the blogosphere saw the removal of this direct censorship policy because it became too onerous to police (see Lee 2010).

Nevertheless, in an era whereby political legitimacy is increasing linked to the morality of governing (Hunt 1999, p. 17), it is not too difficult to understand how and why citizens would overwhelmingly support measures aimed at keeping out harmful and objectionable content, especially when the protection of children or minors are brought into the frame. The ability of the authorities to deny access to selected websites—whether with children or political dissidents in mind—reaffirms the means by which Foucaultian forms of (il)liberal governmentality exercised via the cultural control and conduct of the public sphere, can be enforced and reinforced in Singapore (Lee and Birch 2000). After all, with the Class Licence scheme, any person or group posting content on the vast and ever-enlarging Internet space becomes a de facto licensee as an 'Internet Content Provider', defined under the Class Licence as:

> Any individual in Singapore who provides any programme, for business, political or religious purposes, on the World Wide Web through the Internet; or any corporation or group of individuals (including any association, business, club, company, society, organisation or partnership, whether registrable or incorporated under the laws of Singapore or not) who provides any programme on the World Wide Web through the Internet, and includes any web publisher or any web server administrator.
> (*Broadcasting Act*, IMDA 2013, Clauses 2a and 2b)

The power of a light-touch self-regulatory regime via a Class Licence scheme lies in the fact the scheme operates carte blanche in a 'catch-all' manner, with all Internet Service Providers and Internet Content Providers automatically deemed to be licensed without the need to apply to the IMDA for permission to operate a website or publish online. The ease of setting up personal homepages, individual blogs or participate in citizen journalism and other online publications, including innocuous postings made on social media sites, means that one would become an Internet Content Provider and fall under the jurisdiction of Singapore's Internet policy. Such ambiguously crafted rules widen the scope of policy enforcement, giving the authorities discretionary powers to deal with offenders, or even would-be offenders. In addition, all web providers and users are regularly reminded that all off-line laws and rules also apply to the online world (Ang 2005). Accordingly, the final clause of the Class Licence reads: 'Nothing in this Schedule shall exempt the licensee from complying with the requirements of any other written law relating to the provision of the licensee's service' (IMDA 1996, 2013, Clause 18). With all possible bases covered, the authorities can—and has—openly articulated that Singapore's Internet governance framework has been successful because it operates in a 'light-touch', 'balanced' and pragmatic manner.

The 'automatic' licensing approach of the Class Licence scheme (1996, 2013), coupled with the 'Internet Code of Practice' (1997) and other established laws in Singapore, gives rise to 'auto-regulation', a regulatory discourse where discipline and control is carried out 'automatically' without the need for direct policing or overt surveillance and supervision. The notion of auto-regulation is an appropriation of Foucault's (1977) critique of the disciplinary power of Jeremy Bentham's Panopticon prison structure:

> [T]he major effect of the Panopticon [is] to induce in the inmate a state of conscious and permanent visibility that assures the automatic functioning of power. So to arrange things that the surveillance is permanent in its effects, even if it is discontinuous in its action; that the perfection of power should tend to render its actual exercise unnecessary. [...] It is an important mechanism, for it automises and disindividualises power. (Foucault 1977, pp. 201–202)

Auto-regulation is predicated upon Foucault's belief that power, understood here as the political management of the Singaporean populace, is perfected when it is 'automised' and 'disinvidualised' (Foucault 1977, pp. 201–202). As Foucault points out, the Panopticon architecture provides the principle of how to 'automise' power, and therefore discipline, via a supreme control of one's cultural conduct:

> [T]he principle on which [the Panopticon] was based: at the periphery, and annular building; at the centre, a tower; this tower is pierced with wide windows that open onto the inner side of the ring; the peripheric building is divided into cells, each of which extends the whole width of the building; they have two windows, one on the inside, corresponding to the windows of the tower; the other, on the outside, allows the light to cross the cell from one end to the other. All that is needed, then, is to place a supervisor in a central tower and to shut up in each cell, a madman, a patient, a condemned man, a worker or a schoolboy. [...] The panoptic mechanism arranges spatial unities that makes it possible to see constantly and to recognize immediately. (Foucault 1977, p. 200)

Foucault then goes on to summarize Bentham's principle by declaring that the exercise of power should be 'visible and unverifiable' (Foucault 1977, p. 201). This suggests that surveillance via the policing of citizens— and indeed their Internet sojourns—must be conducted both 'visibly' (or

directly) and 'unverifiably'—that is, indirectly, subtly and/or unobtru-
sively. In an extensive study on Bentham's approach to liberal governmentality via 'indirect legislation', Engelmann points out that the central idea
is to 'enlist the governed as supplementary governors of themselves and
others' (Engelmann 2003, p. 379). Indeed, this same rationale is prop-
agated by Foucault in his liberal governmentality discourse, which relies
on the shaping of the individuals' conduct via cultural control and regulation of oneself (Lee 2010). The aim is to secure conditions by which
individuals will predictably govern themselves and each another in accordance with the social and cultural dictates of order and utility (Engelmann
2003, p. 379).

Bentham's principle of being simultaneously 'visible' and 'unverifiable'
is explained by Foucault thus:

> Visible: the inmate will constantly have before his eyes the tall outline of
> the central tower from which he is spied upon.
>   Unverifiable: the inmate must never know whether he is being looked
> at at any one moment; but he must be sure that he may always be so.
>   The Panopticon is a machine for dissociating the see/being seen dyad:
> in the peripheric ring, one is totally seen, without ever seeing; in the
> central tower, one sees everything without ever being seen. (Foucault
> 1977, pp. 201–202)

More than just surveillance or a policing technology, the Panopticon is
above all a form of governance, or at the very least, a way to produce good
and strong governance. In other words, it is designed for (auto)regulatory
control. The ability to strike a delicate yet dyadic balance between being
visible and unverifiable makes it possible to govern at a distance, where
the only course of 'real' action needed by authorities is to issue regular
regulatory reminders through the media and to fine-tune legislations and
codes from time to time to ensure currency and relevance (Engelmann
2003, p. 374). Indeed, without sounding too clichéd, the way in which
the Singaporean authorities have 'structured' its Internet regulatory laws,
codes and practices—all of which converge to form Singapore's Internet
governance framework—resembles an online Panopticon.

Writing about what he calls the 'politics of comfort' in Singapore,
the media and journalism scholar Cherian George calls Singapore's
tightly consolidated governmental power 'central control' (George 2000).
Although George does not make references to Foucault or Bentham,

his description of the Singapore government's 'central control' mentality mirrors the idea of the supervisory 'central tower' in the Panopticon, a conspicuously privileged position from which to exercise covert power and surveillance on citizens (Foucault 1977, p. 202). The discourse of Internet auto-regulation embodies the key elements of the Panopticon in that one does not know when the 'supervisor', the analogical extension of the authorities, is really watching. As a result, regulation appears to be carried out automatically and with machine-like precision, enabling the disciplining of the Internet economy so that it would deliver strong economic outcomes as well as regulatory compliance for the most part.

CONCLUSION: THE LOCAL IS GLOBAL

According to UNESCO, Internet Governance (IG) is a central issue of global concern in our contemporary era. According to its website:

> Internet governance is the complementary development and application by governments, the private sector, civil society and the technical community, in their respective roles, of shared principles, norms, rules, decision-making procedures, and activities that shape the evolution and use of the Internet.[8]

UNESCO further advocates an "open, transparent and inclusive approach to Internet Governance based on the principle of openness, encompassing the freedom of expression, respect for privacy, universal access and technical interoperability,"[9] which are central tenets of a multi-stakeholder approach adopted by most Western nations. In addition, it espouses ethical use of the Internet that embodies respect for cultural and linguistic diversity in cyberspace. UNESCO's broad understanding of and vision for IG is consistent with and reflects our working definition (expounded in Chapter 1) as *the underlying combination of technological conventions, regulations and systems of management that enable the working of the Internet*. As we have explained, IG is the broader structure upon which Internet regulations—along with rules, codes and operate.

---

[8] UNESCO Internet Governance Forum: https://en.unesco.org/themes/internet-gov ernance (accessed 20 April 2020).

[9] The full statements and declarations of UNESCO's Internet Governance Forum can be found in the website: https://www.intgovforum.org/multilingual/ (accessed 21 April 2020).

But we cannot disregard the China-backed preference for a multilateral model of Internet governance that is both "state-centric and adaptable to evolving conditions" (Lee 2020; see Hong and Goodnight 2020), not least because it is also premised on universal principles of peace, sovereignty, shared governance and shared benefits (Ministry of Foreign Affairs, China, 2017).

In a similar vein, the distinctions between Internet governance and Internet regulation in Singapore are extremely subtle in that the locally-instituted rules, codes and guidelines that form Internet (auto)regulation are also devised to shore up Singapore's global Internet reputation, as well as its economic and political power. In this regard, the local is very much global in the context of Singapore. In other words, Singapore has sought to discipline and control the local Internet space in an 'automised' manner so that it may not only boost its political authority and legitimacy, it may also begin to influence global Internet governance.

This economic mode of geopolitical thinking has been drilled into the Singapore government's psyche since the nation's independence because of Singapore's real and perceived 'vulnerability' as a polity. As Singapore is a small state, referred to condescendingly by former Indonesian President BJ Habibie as a 'little red dot', situated in the middle of the Malay Archipelago—among a historically less-than-friendly neighbour-hood—its global trade and foreign policy has thus been strategically aligned towards forging good relationships with the global big powers, namely USA, China and Russia, whilst remaining sufficiently cordial with its regional neighbours and beyond (Leifer 2000). As the late political scientist Michael Leifer makes clear:

> [T]he foreign policy of Singapore is very much about coping with a vulnerability that has been an abiding concern and theme since an unan-ticipated independence [...]. To that end, Singapore's foreign policy has been informed greatly by the precepts of the balance of power; the eternal goal of such a policy is to deny a hegemonic position on the part of states judges likely to harm the interests of the Republic. (Leifer 2000, p. 9)

Although not expressly spoken of as such, Singapore's internet gover-nance framework is very much premised on its foreign policy thinking which privileges power through economic prowess and the espousal of sovereignty, especially when national security is deemed to be at stake. As a member of the Internet Governance Forum since 2006, Singapore has openly expressed its support of a US-centric multistakeholderism (that is at the core of the Internet Governance). At the same time, however, it has been able to empathise with, even defend, China's push for a state-centred multilateral model because Singapore's light-touch, balanced and pragmatic auto-regulation of the Internet is precisely that: local and state-centric.

While the Chinese government has been accused of attempting to 'hijack' the Internet by advocating for cyber sovereignty in order to export 'digital authoritarianism' (Lee 2020), a charge that remains hotly debated and contested, the Singapore government's internet governance mission appears to be largely, even solely, economic. This is the case even when viewed with global lenses because as a small city-state, Singapore has always articulated its economic ambitions in global terms (Lee 2016). As mentioned in Chapter 3, this can be seen in the way Singapore uses nomenclatures of 'hubs', 'testbeds', 'labs/laboratories' and the likes to frame Singapore as the ultimate destination where innovative ideas and digital technologies can flourish. Seen in this mould, Singapore's Smart Nation initiative is the Internet-centred version of the nation's decades-old 'media hub' ambitions, which is to ensure that the city-state's standing as a central global trade node is not just undiminished, but further augmented (Lee 2016, p. 83).

Jason Luger (2020) puts it cogently in his study of Singapore's extension of its territorial reach via the Internet:

> The implication [...] here is that illiberalism often attributed to, or from, one place (Singapore, in this case) is really multidirectional, non-linear, and often composed of a multitude of global bits and pieces (with some more 'illiberal' aspects originating in oft-perceived 'liberal' places). The application of illiberal state policy or authoritarian restrictions in a specific territory is therefore a combination of local state and society; global state and society; corporations and market factors; as well as the residue of colonial and postcolonial power relations that are repurposed and combined in new ways. (Luger 2020, p. 90)

The key point here is this: because the relationship between liberal and illiberal, or authoritarian, forms of governance is not binary but dialectical, it is important to consider how non-democratic forms of control can in fact liberate societies, if not socially, then economically, which is one way to reconcile Singapore's ability to auto-regulate the Internet without losing its profit-maximising potential (ibid.). Luger issues the reminder that in addition to regular internet users, contemporary government regulators have to grapple with platforms such as a YouTube, Facebook and Twitter—and indeed, the Chinese 'B.A.T.' equivalents of Baidu, Alibaba and Tencent—which functions in some ways like nation-states in their scale, power and autonomous governance structures and by-laws (Luger 2020, pp. 89–90). The dynamic, rapidly changing Internet space is thus likely to lead to conflicting outcomes as democratic governments begin to restrict freedoms in some digital spaces, whilst authoritarian regimes start to open-up new digital forms of communication (as has been in case in many places already). Understood in this way, Internet governance needs to be approached in, as Luger puts it (in the above quote), a multidirectional and non-linear manner.

In deciding to maintain a foreign policy that is acutely aware of its vulnerability as a small state, Singapore's politically neutral stance vis-à-vis the global superpowers of the day in most aspects of international affairs is very much superimposed onto the global Internet governance stage. As a result, Singapore has been walking a tightrope between staying economically connected with new loci of Internet and digital technologies, regardless of political inclinations and affiliations, whilst ensuring that it retains a high-degree of social, cultural and political compliance and control on the domestic front. This makes Singapore not just connected, but hyperconnected and thoroughly hybrid in both political and governance terms.

It also makes the Singapore experience—along with the Malaysian example—an excellent point of departure to think more strategically about the contestations, confluences and consequences of global internet governance that seem to remain in a multistakeholderism versus multilateralism quagmire.

BIBLIOGRAPHY

Ang, P. H. (1999). Information Highways—Policy and Regulation: The Singapore Experience. In Venkat Iyer (Ed.), *Media Regulation for New Times* (pp. 97–114). AMIC: Singapore.

Ang, P. H. (2005). *Ordering Chaos: Regulating the Internet*. Singapore: Thomson Learning.

Australian Broadcasting Authority (ABA). (1997). *The Internet and Some International Regulatory Issues Relating to Content*. A Pilot Comparative Study

Commissioned by the United Nations Educational, Scientific and Cultural Organisation (UNESCO). Sydney: ABA.

Castells, M. (2001). *The Internet Galaxy: Reflections on the Internet, Business and Society*. Oxford: Oxford University Press.

Engelmann, S. G. (2003). "Indirect Legislation": Bentham's Liberal Government. *Polity, XXXV*(3), 369–388.

Foucault, M. (1977). *Discipline and Punish: The Birth of the Prison* (Alan Sheridan, Trans). New York: Random House.

George, C. (2000). *Singapore: The Air-conditioned Nation*. Singapore: Landmark Books.

Hong, Y., & Goodnight, G. T. (2020). How to Think About Cyber Sovereignty: The Case of China. *Chinese Journal of Communication, 13*(1), 8–26.

Hunt, A. (1999). *Governing Morals: A Social History of Moral Regulation*. Cambridge: Cambridge University Press.

Info-Communications Media Development Authority (IMDA). (1996). Broadcasting (Class Licence) Notification. *Broadcasting Act* (Chapter 28, Section 9), Singapore, 15 July.

Info-Communications Media Development Authority (IMDA). (1997). Internet Code of Practice. Singapore, www.imda.gov.sg. Accessed 16 April 2020.

Info-Communications Media Development Authority (IMDA). (2013). Broadcasting (Class Licence) Notification. *Broadcasting Act* (Chapter 28, Section 9), Singapore, Revised 29 May 2013.

Lee, H., & Lee, T. (2019). From Contempt of Court to Fake News: Public Legitimisation and Governance in Mediated Singapore. *Media International Australia, 173*(1), 81–92.

Lee, J. (2020, April 28). Will China Reinvent the Internet? *The China Story*. https://thechinastory.org/will-china-reinvent-the-internet/. Accessed 28 April 2020.

Lee, T. (2002, December). New Regulatory Politics and Communication Technologies in Singapore. *Asia Pacific Media Educator, 13*, 4–25.

Lee, T. (2005). Internet Control and Auto-Regulation in Singapore. *Surveillance and Society, 3*(1), 74–95.

Lee, T. (2010). *The Media, Cultural Control and Government in Singapore*. London and New York: Routledge.

Lee, T. (2016, February). 'Forging an 'Asian' Media Fusion: Singapore as a 21st Century Media Hub. *Media International Australia, 158*(1), 80–89.

Lee, T., & Birch, D. (2000). Internet Regulation in Singapore: A Policy/ing Discourse. *Media International Australia Incorporating Culture and Policy, 95*, 147–169.

Leifer, M. (2000). *Singapore's Foreign Policy: Coping with Vulnerability*. London: Routledge.

Luger, J. (2020). Planetary Illiberalism and the Cyber-City-State: In and Beyond Territory. *Territory, Politics, Governance, 8*(1), 77–94.

Miller, P., & Rose, N. (1990). Governing Economic Life. *Economy and Society, 19*(1), 75–105.

Ministry of Foreign Affairs, People's Republic of China. (2017, March 1). *International Strategy of Cooperation on Cyberspace.* https://www.fmprc.gov.cn/mfa_eng/wjb_663304/zzjg_663340/jks_665232/kjlc_665236/qtwt_665250/t1442390.shtml. Accessed: 29 April 2020.

Yao, S. (1996). The Internet, State Power and Techno-Triumphalism in Singapore. *Media International Australia, 82,* 73–75.

CHAPTER 6

Towards a Hybrid Understanding of Internet Governance: Some Concluding Thoughts

Abstract This chapter explains why a hybrid model of Internet governance that embodies the discourse of glocalisation needs to emerge in order to weaken the binary that Internet governance has been premised upon. We offer some concluding thoughts to the issue, with the key contention that a hybrid model is what both Singapore and Malaysia have been practising since the advent of the Internet, even if such terminologies have not been used. This is mainly because they are seen as small nations, even bit players, in a domain dominated by larger players, namely the US and China. Global internet governance needs to avoid the multistakeholderism versus multilateralism binary to stay nimble, flexible—and also neutral—notwithstanding that libertarian societies will continue to push their own agendas; and, authoritarian societies will continue to assert their right to censor or restrict access to some information, as has been the case throughout media history.

Keywords Hybrid model · Glocalisation · Multistakeholderism · Multilateralism · Liberalism · Neoliberal · Authoritarian · Singapore · Malaysia · China

INTRODUCTION

In many ways this is a volume about in situ Internet governance. The study considers how governance of the Internet responds to rather than lay down the law with its users and their contexts. At the same time, this is a book that seeks to identify the actual or the working model(s) of internet governance that exist(s) between the abstract ideal and the concrete compromise. In analysing the Internet governance practised in Singapore and Malaysia, what we have found is a level of governmental agility particular perhaps to countries that have less truck with liberal democratic history and are more keen to leverage the socio-economic advantages afforded by the Internet. Admittedly, such flexibility is disconcerting for users, both individual and corporate, and have at times been abused for political gains. Indeed, as we have argued in the preceding chapters, both Malaysia and Singapore have approached internet governance with rather similar strategies that largely accept the 'global-ness' of the technology, while insisting that the local, domestic or national elements, including how the Internet is used within its own borders, must be carefully managed. So while the Singapore government has been concurrently praised and lambasted in the past for its purportedly 'liberating light-touch' style of regulation that is 'balanced' with draconian oversight, Malaysia began with a pro-tech, apolitical approach but eventually resorted to extant laws to hinder and prohibit the dissemination of content and information, much of it undesirable because of their critical and political nature.

What we have set out to do in this book is to bring together the various accounts of Malaysia's and Singapore's experience in internet governance. In doing so, we have identified what we refer to as a hybrid model of Internet governance that weakens the binary that Internet governance has been premised upon over more than three decades. We contend that a hybrid model is what both Singapore and Malaysia have been practising, even if they have not used such terminologies to describe their thinking, mainly because they are seen as small nations, even bit players in a domain dominated by larger players. What we call the hybrid model of internet governance here is distinguished by several characteristics. Firstly, it is concurrently neoliberal and authoritarian, or as some would argue, semi-authoritarian with varying doses of market-based capitalism. Secondly, it is adaptable, versatile and responsive to economic and technological change but hostile to socio-political dissent. While Singapore's variant is a combination of hardnosed pragmatism with economic priorities, ethnocratic Malaysia's version is a continual oscillating entanglement with liberalism. Thirdly, and most significantly, this hybrid model is effectively *glocal* in that it recognises that while the technology is always-already global, it will always be used and deployed locally. Adopting a glocal position makes it possible to see internet governance as a global concern yet accept that arguments about sovereignty, or indeed, China's insistence that cyber sovereignty supersedes liberal principles of openness, transparency and inclusivity (Lee 2020). Perhaps the broader question is whether this hybrid model can be emulated beyond Singapore and Malaysia, on a larger scale?

Nations are now grappling with a host of threats that accompany the advances and opportunities that the Internet-enabled flow of words, images, capital, commerce and politicking brings to contemporary societies. How do nations contain the perceived threat to national cyber security yet remain open to cross-border trade in products and services? How do citizens avail themselves of the richness of information being exchanged digitally without being subject to influences of disinformation and malicious rumours from across the world? And, how do we govern the internet when we cannot, with any real confidence, predict which of the thousands upon thousands of technologies, trends and innovations may take off or fall by the wayside of passing fads? Most government are caught between a lack of funds and voter-soothing answers to these challenges. Corporations, however, such as Google and Microsoft have, courtesy of consumer sales and stock values, an abundance of resources to throw at these important societal issues. But realistically, no one can expect these corporations, who have to answer to shareholders, to 'do no evil' if evil includes creating rules that make for better business profits. As Suzor (2019) writes in *Lawless*, we cannot leave the governance of the Internet to the corporations. Extending this rationale into the governmental domain, nor should we leave the governance of the Internet to the strongest or loudest superpower of the day. Letting either side take control of the debate would return the internet to the incompatible binary of liberalism versus authoritarianism, which is ultimately untenable.

Governance has historically been derived from values, ideals that a society more often aspires to than achieved. Compliance with governance indicates the consent of the public to be governed for the greater and common good. Governance has been the remit of national governments rather than markets, technologists or consumers. This is largely because in democracies the assumption is that the government will always work for the good of the majority (of its citizens). What ails global internet governance in our contemporary situation is the lack of agreement over if, which and whose values are universal. Even assuming there is consensus on values worth enshrining in law, there remains the question of who—governments, corporations, individuals, non-profits, global consortiums—should govern the Internet and hence, enforce the regulations. Ironically, it is China, a socialist nation that insists today that internet governance is the right and the responsibility of a country's government and, the West that insists that a global, market-oriented system of Internet governance is the better and established model. With its concept of cyber sovereignty and the runaway success of many of its technological behemoths, China's point of view is gaining traction in the global south. In part, this is a reaction to Western hegemony in these matters and, in part, this is a response to the pace of technological change and the fickleness of user bases. What was valued by the market at millions one day can, in this digital age, tumble down by many zeros next week. The unpredictability of tastes, instability of vaguely worded laws made up as we go along, makes for shorter and shorter boom and bust cycles. As China's technological unicorns race up the value chain, the CCP's approach towards internet governance also becomes more attractive to other nations. In the global south this attraction is tinged with anti-Westernism, a delayed reaction to the colonialism and its lasting legacies across many developing nations, sometimes seen as a means to assert national sovereignty.

At the same time, the ageing populations of mostly Western and developed nations and shrinking disposable incomes have the world's corporations eyeing the rising populations of Asia with varied success. Facebook, for example, is popular in most parts of Asia but has failed to penetrate into the Chinese market. Facebook's Zuckerberg famously wooed the Chinese market, learning Chinese and making repeated visits to China, but to little avail (Kirkpatrick 2018). However, such overtures from the West have, rather than making inroads into new Asian markets, led to the development of domestic tech corporations in Asia. For example, the ridesharing tech disruptor Uber entered the Asian market in 2012, but by 2016 it was quickly outstripped and supplanted by *Didi Chuxing* (or simply DiDi) in China and by Grab in Southeast Asia in 2018 (Russell 2019).

Glocalisation: Hybridising the Global

Much of the talk about globalisation has, almost casually, tended to assume that it is a process which overrides locality, including large-scale locality such as is exhibited in the various ethnic nationalisms which have arisen in various parts of the world in recent years. This tendency neglects two things. First, it neglects the extent to which what is called local is in large degree constructed on a global, or least a pan-or super-local, basis. In other words, much of the promotion of locality is in fact done 'from above.' Much of what appears at first experience to be local is the local expressed in terms of a generalised recipe of locality. Even in cases where there is no concrete recipe – as in the case of some forms of contemporary nationalism – there is still, or so I would claim, a translocal factor at work; the basic idea here being that the assertion of ethnicity and/or nationality is at least made within contemporary global terms of identity and particularity. (Robertson 2012, p. 192)

One of the most influential paradigms for a broad base of disciplines and socio-cultural practices from the 1980s, and through much of the 1990s, was 'globalisation' (Featherstone and Lash 1995). While the term 'globalisation' generally refers to the growing interdependence among the world's societies, it rarely connotes international harmony nor a true global community (Sabo 1993). Rather it recognises that what happens in a society is influenced by events and experiences in another part of the world. This is perhaps why Stuart Hall (1992) has argued that "global and local are the two faces of the same movement" that we refer to as glob-alisation. For many scholars, globalisation is problematic and ambiguous because it is often implicated in a process of cultural and political trans-formation and dependency. What most people refer to as globalisation when relating to social, cultural, economic and technological progress often draw inspiration from just a "few power centers" (Thompson 1995, p. 166). In the 1990s, this centre was very much the communication and cultural commodity production base of the United States of America (US), acting as proxy for the 'West'. This was extended during the early days of public Internet access, in the 1990s especially, as this was very much dominated by the Anglo-American 'West'. As a result, globalisation studies expanded into the realm of culture as concerns shifted to notions of 'cultural homogenisation' (Hall 1992), 'cultural imperialism' (Tomlinson 1991), and also 'cultural colonisation' with its various popularised terms such as 'McDonaldization' and 'Coca-Colonisation' (Ritzer 1993).

Roland Robertson, one of the key globalisation scholars, argued that the globalisation phenomenon of the contemporary era saw the world becoming united, not integrated (1990, p. 18). This was because people would instinctively seek to apply any global discourse to their immediate local context and situation in order to make it directly relevant to their particular circumstance (as exemplified in the opening quote to this section). Robertson also opposes the idea that all forms of locality are substantively homogenised when confronted by globalisation (1995, p. 31). Consequently he advocates the notion of 'glocalisation' as a refinement of the concept of globalisation (Robertson 2012). The idea of glocalisation is modelled on the Japanese word *dochakuka*, which is the agricultural principle of adapting one's farming technique to local conditions (ibid.). In the business world, this idea was adopted to refer to "global localisation" (Khondker 2005), which means to consider the local or one's own localised social or cultural practice as a micro manifestation of a global variety. Robertson's glocalisation thesis offers us a way of understanding global Internet governance because in the process of either embracing or rejecting globalisation and global practices, along with the many criticisms that accompany the discourse, especially those relating to questions of authenticity, it is easy to forget that local is a subset of the global. Indeed, forms of social categories and practices would typically assume local flavours or characters despite the fact that they may be created or invented elsewhere (Khondker 2005, p. 185).

In his extensive study on the shift from globalisation to glocalisation, sociologist Habibal Haque Khondker (2005) observes that the discourse has been referred to in other disciplines and in other linguistic contexts, as *melange, syncretism* and, significantly, *hybridity*. It is important however to point out that glocalisation and hybridity are not the same thing. Glocalisation involves blending, mixing and adapting two or more processes of which one must be local or contain local ingredients. The concept of hybridity however does not necessarily involve local elements, and

it can be made up of a multiplicity of glocal ideas and global practices (Khondker 2005, p. 191). We can argue that a hybrid idea can be
glocalised, applied and made to work in different contexts.

Understood in this way, the quest to identify and institute a common
structure for global Internet governance is very much an attempt to
hybridise the global. In many ways, this is achievable yet seems insur-
mountable because the stakeholders embody different points of departure. This is perhaps where the problem resides: proponents for both
the multistakeholderist and the multilateralist approaches have not yet
grasped the need to adopt a hybrid model that takes into account the
fact that the Internet is not merely a global technology; it is also thoroughly glocal, both ideologically and in practical terms. If we begin to
take a glocalised/hybrid view of where global internet governance ought
to be, we immediately bring back and emphasise the importance of a
'locality'—in that the vast majority of Internet use and sojourns occur at
local levels—that is enmeshed or infused with the 'global' domain. The
global dimension of the Internet thus enable us to connect, interact and,
in the process, enrich and hybridise our contents and ideas with those
from another locale and temporality.

Conclusion

It is worth reiterating in closing here that the main premise of this book
is to propose a hybrid model of Internet governance—and as we have
emphasised, it is one that already exists. Internet studies and governance
scholars who have observed and participated in the debate over many
years would do well to look more closely at the Internet regulatory and
governance approaches that Malaysia and Singapore have been developing
and fine-tuning since the early 1990s. Regulators and government officials
from Malaysia and Singapore could also begin to share information and
data about how their governance strategies have been applied, including
policy and technical issues that are to be improved or calibrated, especially
as new technological challenges emerge. Yet this is not without problems.
The most challenging of course is that any mention of direct or indirect
censorship, or auto-regulatory structures of surveillance and controls via a
suite of illiberal laws (such as internal security, sedition or anti-fake news
laws that remain very much in force in both Malaysia and Singapore)
would put off countries that are constitutionally-bound by or steeped in
liberal democratic practices.

At the same time, we cannot discount the relative economic and commercial successes that both Malaysia and Singapore have consistently reaped in their embrace of the digital sphere, with their respectively digital economy valued at approximately RM270 million for Malaysia in 2019 (Heng 2020), and forecast to reach SGD13.5 billion in Singapore by 2021 (Tang 2020). Singapore's success in the economic realm—with a 'third world to first world in a generation' rhetoric that was made famous by her founding Prime Minister, the late Lee Kuan Yew—has found many global admirers from different countries, of different political stripes and outfits. As co-author Terence Lee has noted in another study, a number of countries in Asia had already begun studying Singapore's Internet policies and regulatory model back in the early-2000s, with an intent to emulate aspects of its economically-favourable control methods (Lee 2004).

Veritably, we recognise that our approach is unconventional because in the matrix of global affairs, Southeast Asia is often merely the passive theatre where major powers preen and battle for supremacy. Even in histories of the internet, few accounts speak of the pivotal role that lowered costs, courtesy of the millions producing personal computers, integrated circuits and mobile phones at factories across Asia, play in the popularisation of the World Wide Web, instant messaging or indeed, of the ubiquitous yet humble email. In fact, apart from the rare studies on factory labour in China (Qiu 2016), there are still few studies that mention Southeast Asian labour, let alone write of internet governance in the region as worthy of study beyond critique. This scholarly gap calls out for analysis, which is but one which this book has sought to fill.

We have shown how Malaysia and Singapore have both led the way in entrenching the economic lure of technological innovations in the broader region. If this extension of an industrial mindset has turned Southeast Asia into a testbed for the powers-that-be, then its developments deserve to be studied on their merits. In our chapters, we have worked in discussions of how the legacy of British era media infrastructure and regulations equipped postcolonial nations, Malaysia and Singapore, to cope with the unexpected and explosive mass popularity and indispensability of the Internet. We also examined the factors, events and aspirations that motivate the strategies advanced and policies adopted in the three decades (from the 1990s) that the Internet was introduced to the region. While we expect tensions surrounding global internet governance to continue beyond the publication of this book, we believe that by there is much to glean by consolidating the hybrid experiences and influences that we have observed and studied in Malaysia and Singapore over a protracted period of time.

This is important because it is very likely that global Internet governance will have to account for both multistakeholderist and multilateral ideologies and practices for some time yet. As the world continues to recover from the COVID-19 pandemic that struck in 2020, which has forced a global retreat into new forms of protectionism and diplomatic conflicts, the re-enactment of trade and human movement borders would shore up state-driven sovereignty. Along with this shift will be a greater acceptance of the importance of cyber or Internet sovereignty, which will give China an immediate boost to its preference for a multilateral approach to governance. But to view this outcome via triumphalist lenses would be a mistake because most nations would also opt to remain under the multistakeholderism umbrella because, like it or not, the Internet is inherently global. The rationale for supporting a multilateral discourse that promotes nationalised networks that remains connected to the broader world wide web would be to enable governments to forge new bilateral, multilateral or regional ties to navigate their countries back to prosperity in a post-COVID-19 world. While immediate needs such as food sufficiency and safety, medical equipment, technologies and other essential supplies are top priorities, Internet access, reliability and digital security to ensure the economy continues to function well are a close secondary priority (Sathirathai 2020).

This global shift does not mean that countries like China will not attempt to push its cyber sovereignty agenda more strongly to take advantage of a seemingly fragmented global order or indeed a disjointed digital space (Dupont 2020). On the contrary, China will double its efforts to politicise the Internet governance discourse and convince its allies and supporters, including the ASEAN bloc of nations, to support its vision of an alternative internet protocol that allows for multilateral governance to operate—and also allow the Chinese big tech companies economic in-roads into new markets. By the same token, the US and its allies who support the free flow of the Internet that underpins the multistakeholderism model would campaign hard to ensure that the Internet does not get 'Balkanised' by the Chinese. As Alan Dupont, an Australian geopolitics specialist and professor of International Security, warns quite explicitly:

A fractured digital world would vastly complicate e-commerce and trade, restrict the free flow of information, reduce international collaboration and human interaction, and leave us vulnerable to the exploitation of our [present] relatively open system by authoritarian states secure behind their digital firewalls. To prevent these outcomes, the government will need to engage with friends and allies to come up with a fit-for-purpose worldwide web that is more efficient, secure, user friendly and compatible with democracy. (Dupont 2020)

In short, while the post-COVID-19 era will show up new local, regional and global urgencies and priorities, the debate and haggling over global Internet governance will continue unabated—or it may even intensify.

What all this points to is that global internet governance will have to avoid the multistakeholderism versus multilateralism binary position in order to stay sufficiently nimble, flexible—and to an extent, neutral, notwithstanding that authoritarian-leaning societies will continue to assert their right to censor or restrict access to some content and information, as has been the case throughout media history. Writing of the decolonization of internet governance Syed Mustafa Ali (2018) argues that "framing the issue [of internet governance] in terms of universality versus *enclosure* involves recourse to the historical experience of European feudalism while obscuring historical colonialism and the persistence of racialized coloniality in core-periphery relations" (p. 143). He suggests instead that "it might be *necessary* to oppose a commitment to Internet 'universality' and 'openness' in favour of statist alignment as a temporary tactical manoeuvre within a strategic decolonial 'horizon'" (p. 170).

It is our contention that internet governance as practised by Malaysia and Singapore, speaking somewhat as proxies for other Southeast Asian states, presents precisely such an option—or indeed a model that embodies elements of being glocal in its very hybridity. As global Internet governance enters a new phase that will be referred to for some time as a post-COVID-19 era, these smaller Southeast Asian states should begin to articulate, with confidence, their successful experiences of a hybrid Internet governance model.

BIBLIOGRAPHY

Ali, S. M. (2018). Prolegomenon to the Decolonization of Internet Governance. In D. Oppermann (Ed.), *Internet Governance in the Global South: History, Theory and Contemporary Debates* (pp. 109–183). São Paulo, Brazil:

International Relations Research Center, Núcleo de Pesquisa em Relações Internacionais (NUPRI), University of São Paulo.

Dupont, A. (2020, May 16). China's Bid to Control the Internet. *The Australian*. https://www.theaustralian.com.au/inquirer/china-sets-out-a-snare-for-the-worldwide-web/news-story/8b731f74923842bbc428436cf43 e950d. Accessed 16 May 2020.

Featherstone, M., & Lash, S. (1995). Globalization, Modernity and the Spatialization of Social Theory: An Introduction. In M. Featherstone, S. Lash, & R. Robertson (Eds.), *Global Modernities* (pp. 1–24). London: Sage.

Hall, S. (1992). The Question of Cultural Identity. In S. Hall, D. Held, & T. McGrew (Eds.), *Modernity and Its Futures* (pp. 274–316). Cambridge: Polity Press.

Heng, D. (2020, January 25). Malaysia Sees the Digital Economy as Key to Unlocking Rural Economic Growth. *ASEAN Today*. https://www.aseant oday.com/2020/01/malaysia-sees-the-digital-economy-as-key-to-unlocking-rural-economic-growth/. Accessed 1 June 2020.

Khondker, H. H. (2005). Globalisation to Glocalisation: A Conceptual Exploration. *Intellectual Discourse, 13*(2), 181–199.

Kirkpatrick, D. (2018). Facebook Is Bent on Friending China, but .... *South China Morning Post*. https://www.scmp.com/week-asia/business/article/2159102/facebook-bent-friending-china. Accessed 22 June 2020.

Lee, T. (2004). Emulating Singapore: Towards a Model for Internet Regulation in Asia. In S. Gan, J. Gomez, & U. Johannen (Eds.), *Asian Cyberactivism: Freedom of Expression and Media Censorship* (pp. 162–196). Bangkok: Friedrich Naumann Foundation.

Lee, J. (2020). Will China reinvent the Internet? *The China Story*, 28 April. https://thechinastory.org/will-china-reinvent-the-internet/. Accessed 28 April 2020.

Ritzer, G. (1993). *The McDonaldization of Society*. Thousand Oaks, CA: Pine Forge Press.

Robertson, R. (1990). Mapping the Global Condition: Globalization as the Central Concept. In M. Featherstone (Ed.), *Global Culture: Nationalism, Globalization and Modernity* (pp. 15–30). London: Sage.

Robertson, R. (1995). Glocalization: Time-Space and Homogeneity-Hetereogeneity. In M. Featherstone, S. Lash, & R. Robertson (Eds.), *Global Modernities* (pp. 25–44). London: Sage.

Robertson, R. (2012). Globalisation or Glocalisation? *Journal of International Communication, 18*(2), 191–208.

Russell, J. (2019). Uber Has Already Made Millions From Its Exits in China, Russia and Southeast Asia. *TechCrunch*. https://techcrunch.com/2019/04/11/uber-global-exits-billions/. Accessed 22 June 2020.

Sabo, D. (1993). Sociology of Sport and New World Order. *Sports Science Review*, 2(1), 1–9.

Sathirathai, S. (2020, May 6). Challenges and Opportunities for S-E Asia in Post-Covid-19 World. *The Straits Times*. https://www.straitstimes.com/opinion/challenges-and-opportunities-for-s-e-asia-in-post-covid-19-world. Accessed 6 May 2020.

Suzor, N. (2019). *Lawless: The Secret Rules That Govern Our Digital Lives*. Cambridge: Cambridge University Press.

Tang, S. K. (2020, January 13). The Rise of the Digital Economy: What Is It and Why It Matters for Singapore. *Channel News Asia*. https://www.channelnewsasia.com/news/business/what-is-digital-economy-why-it-matters-mobile-app-12240630#:~:text=By%20one%20estimate%2C%20the%20burgeoning,research%20firm%20IDC%20Asia%2DPacific. Accessed 1 June 2020.

Thompson, J. B. (1995). *The Media and Modernity: A Social Theory of the Media*. Cambridge: Polity Press.

Tomlinson, J. (1991). *Cultural Imperialism*. London: Pinter Publishers.

Qiu, J. L. (2016). *Goodbye iSlave: A Manifesto for Digital Abolition*. Champaign: University of Ilinois Press.

INDEX

A

Ang, P.H., 36, 39, 72, 73, 82
Anti-Fake News Act 2018, 57
Anti-Westernism, 14, 20, 59, 60, 65, 66, 94
Australian Broadcasting Authority (ABA), 72, 78
Australian Communications and Media Authority (ACMA), 72, 77
Authoritarian/authoritarianism, 35, 36, 38, 44, 61, 73, 75, 81, 87, 88, 92, 93, 100
Auto-regulation, 44, 74–77, 80, 83, 85, 87

B

Balanced and light-touch, 74, 78
Balkanisation, 5
Belt and Road Initiative (BRI), 57, 59, 65
Bentham, Jeremy, 83, 84
Bersih (clean) rally, 24, 53

C

Capitalism, 25, 59, 61, 79, 92
China, 2, 4, 5, 7, 27, 42, 46, 56–62, 65, 66, 86, 87, 93–95, 98, 99
Class Licence scheme, 74, 78, 82, 83
Colonialism, 52, 94, 100
Cyberlibertarian, 17, 18
Cyber Security Agency of Singapore (CSA), 43
Cybersecurity Law of China, 61
Cyber sovereignty, 87, 93, 94, 99
Cybertroopers, 53
Cyber wellness, 79, 80

D

DeNardis, L., 3
Digital authoritarianism, 87
Digital Free Trade Zone (DFTZ), 57, 58
Digital Government Office (DGO), 33, 34, 43, 47
Discipline, 14, 46, 76, 83, 86, 95, 96
Disinformation, 56, 93

E

Electronic World Trade Platform (eWTP), 58
European Union (EU), 3, 63
Expedience, 52

F

Foucault, Michel, 44, 75, 76, 82–85

G

General Data Protection Regulation (GDPR), 3, 4, 63
Global digital economy, 47
Global economic hub, 47
Globalisation, 47, 65, 95, 96
Glocal, 93
Glocalisation, 95, 96
Governmentality, 34, 75, 76, 82, 84
Government Technology Agency (GovTech), 42, 43

H

Hall, Stuart, 95, 96
Hindraf Rally, 24
Hybrid configurations, 5, 6
Hybridising, 25, 95
Hybrid model, 14, 52, 66, 75, 92, 93, 97

I

Industrial technology, 21
Infocomm Development Authority (IDA), 33, 39–43
Infocomm Media Development Authority (iMDA), 43, 76–79, 82
Information and Communication Technologies (ICTs), 20, 32, 34, 37, 41
'Intelligent Island', 32, 33, 36, 37, 40

Intelligent Nation 2015 (iN2015), 33, 39, 40
Internet Governance Forum (IGF), 5, 86
Internet Governance (IG), 2, 3, 5–9, 14, 26, 27, 33, 37–39, 43, 47, 48, 52, 54, 56, 60–65, 72–78, 80, 82, 84–88, 92–94, 96–100
define, 8
Internet self-regulation, 79
Internet sovereignty, 39, 60–62, 65, 73, 99
Interoperability, 4, 85

L

Lee, Hsien Loong, 34, 35, 45
Lee, T., 35–40, 42, 44, 73, 74, 76, 80, 81, 84, 98
Leifer, Michael, 86
Liberalism, 2, 38, 61, 93
'Little red dot', 86
'Look East' policy, 14, 16, 26

M

Mahathir, Mohamad, 7, 13, 14, 16–23, 25–27, 55, 59, 60, 62
Vision 2020: the Way Forward, 16, 17
Mainstream and broadcasting media, 22
1Malaysia, 53
Malaysia Communications and Multimedia Commission (MCMC), 55
Malaysian Institute of Microelectronic Systems (MIMOS), 19
Media Development Authority (MDA), 33, 77, 78, 80
Media governmentality, 36
Media literacy, 56, 79, 80

MSC Malaysia Bill of Guarantees
    (BoGs), 20, 21
Mueller, M., 2, 61
Multilaterialism, 2
Multimedia Super Corridor (MSC),
    16, 18–22, 25, 26, 47, 54, 55,
    58, 59, 62
Multistakeholderism, 2, 75, 87, 88,
    99, 100

N
National cyber security, 43, 93
Neoliberal, 76, 92

P
Panopticon, 83–85
Platform governance, 5, 63
Post-COVID-19, 99, 100
Protection from Online Falsehoods
    and Manipulation Act (POFMA),
    75
Public oversight, 62
Public surveillance, 34

R
Radu, R., 5, 6
Regulation, 3, 6–8, 16, 20, 21, 34,
    55, 63, 72–74, 78, 85, 86, 94,
    98
Regulatory compliance, 85
Robertson, Roland, 95, 96

S
Scale, 3, 8, 21, 34, 57, 59, 88, 93
Semi-authoritarian, 24, 52, 61, 92
Singapore Broadcasting Authority
    (SBA), 41, 77, 78, 81
Singapore Internet Project (SIP), 40,
    41
"Smart Nation", 33–35, 42, 43,
    45–47, 74, 87
Socio-historical approach, 9, 52
Standards, 4, 5, 41
Surveillance, 27, 44, 74, 75, 83–85,
    97
    of technology, 77
Suzor, N.P., 5, 63, 64, 93

T
Tan, K.P., 35, 65
Technology corridor, 19
Techno-nationalism, 6
'Test-bed', 46
Test drive, 18

U
United Nations Educational, Scientific
    and Cultural Organisation
    (UNESCO), 72, 85
Urban digitalisation, 34

V
A Vision of an Intelligent Island:
    IT2000 Report (1992), 33, 37